Giraffe

Biology, Behaviour and Conservation

With its iconic appearance and historic popular appeal, the giraffe is the world's tallest living terrestrial animal and the largest ruminant. Recent years have seen much-needed new research undertaken to improve our understanding of this unique animal.

Drawing together the latest research into one resource, this is a detailed exploration of current knowledge on the biology, behaviour and conservation needs of giraffe. Dagg highlights striking new data, covering topics such as classification of races; the apparent role of infrasound in communication; biological responses to external temperature changes; and motherly behaviour and grief. The book discusses research into behaviour alongside practical information on captive giraffe, including diet, stereotypical behaviour, ailments and parasites, covering both problems and potential solutions associated with zoo giraffe. With the giraffe becoming an endangered species in Africa, the book ultimately focuses on conservation measures to halt population decline.

Anne Innis Dagg, PhD, teaches in the Independent Studies Program at the University of Waterloo, Canada. Her passion for giraffe has inspired much of her research. In 1976 she co-authored the first scientific book on the species, and in 2010 she was honoured for her pioneering work at the inaugural meeting of the International Association of Giraffe Care Professionals. She is also the author of *Animal Friendships* (Cambridge, 2011).

Giraffe

Biology, Behaviour and Conservation

ANNE INNIS DAGG

Independent Studies Program, University of Waterloo, Ontario

CAMBRIDGE
UNIVERSITY PRESS

University Printing House, Cambridge CB2 8BS, United Kingdom

One Liberty Plaza, 20th Floor, New York, NY 10006, USA

477 Williamstown Road, Port Melbourne, VIC 3207, Australia

314-321, 3rd Floor, Plot 3, Splendor Forum, Jasola District Centre, New Delhi - 110025, India

79 Anson Road, #06-04/06, Singapore 079906

Cambridge University Press is part of the University of Cambridge.

It furthers the University's mission by disseminating knowledge in the pursuit of education, learning and research at the highest international levels of excellence.

www.cambridge.org
Information on this title: www.cambridge.org/9781107610170

© A. I. Dagg 2014

First published 2014
3rd printing 2015
First paperback edition 2019

A catalogue record for this publication is available from the British Library

Library of Congress Cataloging in Publication data
Dagg, Anne Innis.
Giraffe : biology, behaviour, and conservation / Anne Innis Dagg, Independent Studies Program, University of Waterloo, Ontario.
    pages   cm
Includes bibliographical references and index.
ISBN 978-1-107-03486-0 (alk. paper).
1. Giraffe. I. Title.
QL737.U56D27   2014
599.638–dc23   2013040918

ISBN 978-1-107-03486-0 Hardback
ISBN 978-1-107-61017-0 Paperback

For Julian Fennessy, Zoe Muller and Paul Rose
who are devoting their lives to the benefit of giraffe and
who have helped greatly in the writing of this book

Contents

Preface

It all began when, as a toddler, I saw the giraffe at the Brookfield Zoo in Chicago. I was so captivated that I later studied biology at the University of Toronto, hoping to learn everything about the species. This didn't happen – there was no interest then in Africa or in animal behaviour in academic biology. After graduating, my aim was to go to Africa to study giraffe as soon as possible, but I had no contacts there to make this happen. I decided instead to do graduate work for a Master's degree at the university while I wrote letters to see who might help me accomplish my dream. This took many months – letters to government officials or wildlife departments in countries where there were giraffe, letters to names of people dredged up by friends, letters to professors connected with Africa, even letters to L. S. B. Leakey who was to launch Jane Goodall on her career five years later. After early rebuffs I used initials for my signature so the recipient would presume I was a man, but this did not help.

Luckily, about that time Rufus (C. S.) Churcher came from Africa to earn his doctorate at the University of Toronto; he would go on to become a professor there and author of a definitive work on fossil giraffe, 'Giraffidae' (1978). He told me about a professor he had studied with, Jakes Ewer of Rhodes University in Grahamstown, South Africa, who might be able to help me. Jakes and his wife, Griff Ewer, were both willing to do this. They put me in touch with Alexander Matthew who managed a citrus and cattle ranch near the Kruger National Park on which roamed nearly 100 giraffe; after some hesitation – he had assumed I was a man – he finally agreed to have me live and work at his ranch. These amazing people became friends of mine for life.

As the first person to study giraffe in the field, in 1956–1957, I set myself a huge agenda. No one had ever before researched a wild animal like this in Africa, so I had no reason to believe that anyone would be interested in studying giraffe in the future, either. It had taken half my money and nearly a month's travel by ship and a 1000-mile trip in a second-hand car to reach the giraffe in what was formerly the Transvaal, so I knew few people could afford the expense and time to do this. Planes were so pricey they were off my radar. Who could ever afford them? (Foretelling the future is not my strong point, needless to say.) Because of these huge obstacles I felt I had to learn as much about giraffe as I could in the year I had at my disposal before my money ran out (Dagg, 2006). I noted what plants they ate, where they roamed, who seemed friendly with whom, where and when they drank, males who 'necked' with other males, and the anatomy of a dead giraffe. The results are described in my first scientific paper (Innis, 1958).

Back in Canada, I used the footage of giraffe movements that I had filmed in South Africa to earn my PhD with a comparison of the giraffe's gaits with those of other large ungulates. I also gathered material for a manuscript which was to include everything that had been written on this species and, on my husband's sabbatical, researched the behaviour of the 18 giraffe in the Taronga Zoo of Sydney, Australia.

While studying biology at the University of Toronto, one of my classmates was Bristol Foster, who would go on in the 1960s to study giraffe for three years in the Nairobi National Park. He collected valuable information on their distribution and social behaviour. I eventually co-authored with Foster *The Giraffe: Its Biology, Behavior, and Ecology* (1976), which I updated in 1982.

After decades of trying unsuccessfully to become a university professor (it was difficult being a woman scientist at the time, futilely attempting to break into male university departments of biology or zoology), I wrote various books about animals and, in 2006, the story of my year in Africa, *Pursuing Giraffe: A 1950s Adventure*. This led to interviews for the hour-long CBC radio programme 'Ideas' (available on my website annedagg.ca), which in turn led to my signing movie rights about the adventure for a possible movie.

In 2010, I was thrilled to be invited to and honoured at the inaugural conference of the International Association of Giraffe Care Professionals (IAGCP) founded by Amy Phelps of the Oakland Zoo and Paige McNickle and Lanny Brown of the Phoenix Zoo. There I met and talked to many of the 150 attendants, all interested in giraffe research in general, and most involved with the care of giraffe in zoos. They were an added impetus to write this present book.

***

Giraffe: Biology, Behaviour and Conservation (2014) is unlike our first book on giraffe. For it, Foster and I scraped together everything that had been written about this species along with what we had learned in our own researches. Since then, in the past 30 years, scores of zoologists and biologists have produced hundreds of scientific papers on new discoveries about this wonderful animal carried out in the wild, in zoos and in laboratories. While our earlier work was wanting because of a paucity of material, this book feeds on the flood of new discoveries. This cutting-edge research is the basis for this book.

Sadly and paradoxically, while research has flourished, giraffe have not. During the past 15 years, because of burgeoning human population pressures in African countries, illegal killings and drought, the numbers of giraffe in Africa have plummeted by 40%. Unless something drastic is done to stop this catastrophic decline, giraffe could become extinct in the wild. This book therefore emphasizes conservation.

Acknowledgements

The author is most grateful to the many giraffe lovers who have provided information and/or photographs for this book. They include:

Janos Aczel, Karl Ammann, Lance Aubrey, Meredith Bashaw, Douglas Bolger, Mike Bona, Ana Bordjan, Rick Brenneman, David Brown, Lanny Brown, Gary Bruce, Sara Brunton, Alan Cairns, Linda Carson, Kerryn Carter, Lisa Clifton-Bumpass, Ian Innis Dagg, Sharon Dahmer, John Doherty, Tom East, Katie Ellis, Julian Fennessy, Stephanie Fennessy, Christine Filipowicz, Katherine Forsythe, Bristol Foster, Susan Gow, Stuart Green, Ashleigh Kendrac, Vaughan Langman, Lindsey Long, Andri Marais, Paige McNickle, Strahinja Medi, Anne Miner, Zoe Muller, Aimee Fannon Nelson, Ken O'Driscoll, Jack Pasternak, Dale Peterson, Amy Phelps, Jason Pootoolal, Amy Roberts, Sarah Roffe, Paul Rose, Russell Seymour, Florian Sicks, Megan Strauss, Kathy Szigeti, Megan Waddington and Kristen Wolfe.

A special thanks to those who vetted chapters or parts of chapters before the publication of this book: David Brown, Julian Fennessy, Vaughan Langman, Graham Mitchell, Paul Rose and Michael Schlegel.

Abbreviations

BU	Browse Unit
DB	Dallol Bosso
GED	Giraffes Ear Disease
GSD	Giraffe Skin Disease
HIF	human-impact factor
IAGCP	International Association of Giraffe Care Professionals
IM	Intermediate Zone
KNP	Kruger National Park
LNNP	Lake Nakuru National Park
NNP	Nairobi National Park
REM	rapid eye movement
SFA	Serous Fat Atrophy
TST	total sleep time

1   Time-line of giraffe

This chapter has four parts. First, a discussion of the contentious topic of why giraffe evolved to have a long neck. Was it for food? Was it for sex? Next is a short section on the ancestors of giraffe which moved into Africa from Asia millions of years ago and eventually died out there. Then a discussion of the DNA of *Giraffa camelopardalis* itself and what it tells us about the numbers (contested) of subspecies or races spread out over most parts of Africa, information expanded in Chapter 11 on conservation. The chapter ends with an extended description of the giraffe's place in European history.

Evolution – why are giraffe so tall?

Teachers often use giraffe as a tool to explain conflicting theories of evolution. Lamarck believed that the height of giraffe was caused by the acquired characteristics of its forebears; calves had longer necks because their parents and ancestors had acquired them by reaching up continually to obtain browse to eat. By contrast, zoologists today believe in the selection of natural traits. Changes such as longer necks began by chance mutations in individual giraffids. These individuals were more successful than their shorter-necked friends in that more of their progeny survived to breed themselves. Over millennia the necks of the ancestors of giraffe gradually elongated.

Now teachers use giraffe for a new evolutionary lesson. Did giraffe evolve long necks so that they could obtain more food? Or did they do so because females chose to mate with large males who had the longest necks? The obvious reason is that their long necks and legs enable them to browse at heights other large herbivores cannot reach. Elissa Cameron and Johan du Toit (2007) favour this explanation because they found that in Kruger National Park, giraffe did indeed feed on favoured tall trees such as *Acacia nigrescens*. They ingested more leaf mass per bite high in trees than they did when browsing on lower bushes (and in consequence left more food for other ungulates). (Or, conversely, they ate high leaves because the lower ones were already being consumed by these smaller animals.)

However, Robert Simmons and Lue Scheepers (1996) disagree with this idea. They noted that even in the dry season when feeding competition among all ungulates is most intense, giraffe generally fed on low shrubs rather than tall trees. Females spent most of their feeding time with their necks horizontal, in which position both sexes fed fastest. Indeed, browsing by giraffe on low plants rather than tall trees is common in many parts of Africa (Dagg and Foster, 1982). Simmons and Scheepers were the first to suggest that giraffe height did not evolve because of feeding competition, but because of sexual selection. Male giraffe fight over the chance to mate with females in oestrus by clubbing each other with their heads; these become a highly effective weapon when mounted on a long neck. The winning animal gets to pass his DNA on to the next generation.[1]

In response to this suggestion, Graham Mitchell and his colleagues (2009a and 2013) argue instead that from measurements of many dead giraffe bodies, they have found that the growth patterns of male and female heads and necks are similar, and 'that sexual dimorphism of the head and neck is minimal and can be attributed to secretion of sex steroids.'

Food or sex? Robert Simmons, elaborating on his earlier paper but this time with R. Altwegg (2010), examined these two main competing hypotheses. For the first hypothesis, as we have seen, early giraffids with longer necks than those of other species would be better able to survive food bottlenecks during dry seasons to pass on their genes to the next generation. However, giraffe often forage or browse at shoulder height during winter shortages. As well, giraffe are much taller than they need to be to outreach other animals, and have been for more than a million years. Why would they have evolved to a height beyond, say, 10 feet, which would be enough to out-compete all other ungulates? (Elephants and rhinoceros forage on the leaves of trees they knock down, but their behaviour would not compete with that of giraffe.) We must keep in mind, though, that the various species of giraffids evolved when large herbivorous animals such as deinotheres, mastodonts, sivatheres and baluchitheres, now extinct, were competitors for browse.

The second necks-for-sex hypothesis assumes that over time males evolved increasingly longer necks so they could more effectively combat other males for the right to mate with females in oestrus. Females don't fight, but two males do after positioning themselves side by side, whacking each other by swinging their necks to club their opponent's body with their heavy head and horns (ossicones). Rarely is a male killed in such a battle.

There is ample evidence to support this possibility:

(a) males with larger necks are more likely to win combats and to mate with females (Pratt and Anderson, 1982, 1985);

(b) females in oestrus prefer to mate with large-necked and older males – neck mass increases with age in males but not in females (Pratt and Anderson, 1985);

(c) males in dominance contests have a display walk during which they hold their heads high, indicating height is a measure of dominance (Innis, 1958);

(d) skulls of adult male giraffe weigh considerably more than those of adult females – about 10 kg vs. 3.5 kg (Dagg, 1965); and

(e) in theory, sexually selected features of males are often correlated with negative costs for them, which is true for giraffe (Kodric-Brown and Brown, 1984). For example, males are more often killed by lions than are females, which may be related to their often being

---

[1] This original concept of sexual selection was argued by P. Senter (2006) to apply to sauropod dinosaurs which also had a very long neck. However, as these animals are all extinct, there is no way to prove this.

alone while searching for females in oestrus (Owen-Smith, 2008). Their larger size and thus need of more rich browse may also disadvantage them if food sources dry up.

In early fossil giraffids without long necks such as *Giraffokeryx* sp., the horns were on the side of the head and backward-pointing (Churcher, 1978). Their positioning would indicate that males fought with sideways motions of the head, as okapi also seem to, perhaps standing side by side as do giraffe. During their evolution the necks of ancestral giraffids increased in length and mass while the horns became shorter and more compact. With such weaponry they would have fought using their heads as clubs (as opposed to forward-facing horns which imply head-butting (mountain sheep) and antlered heads which imply head wrestling (elk and moose)). Ludo Badlangana, Justin Adams and Paul Manger (2009) speculate how the long neck of giraffids may have evolved beginning about 14 to 12 million years ago.

The unique feature of the giraffe, its neck, is so spectacular that these same researchers also delved into how it differs anatomically from those of 10 other artiodactyls. Although the giraffe neck is of course much longer, comprising over half the length of the vertebral column, it is also composed of seven elongated cervical vertebrae each scaled appropriately for that particular vertebra. (Although an eighth vertebra has been suggested, it does not in fact exist (Solounias, 1999).) This scaling is also characteristic of the cervical vertebrae present in fossil giraffes. The related okapi does not have a significantly long neck, so its lineage separated from that of the giraffe before 12–10 million years ago; it was during the late-middle Miocene that the necks of some giraffids began elongating. The researchers hypothesize about how these changes in the neck lengths of giraffids may have occurred, which was different than how the necks of the camel and llama or the gerenuk lineages evolved.[2,3]

Fossil remains of African giraffids

Members of the giraffe family, Giraffidae, were and are medium to large ruminating artiodactyls, distinctive because almost all have ossicones on their heads which resemble horns, but which are covered by vascularized skin. (In this book I shall call these prominences of giraffe 'horns', as most authors do, despite their difference from true horns that are present in cattle.) Giraffids also had in common large sinuses below these horns, and lower canines that were bilobed and therefore wider and flatter than those in most species. Strangely, though, they did not have long necks like the giraffe; that was a relatively recent adaptation perfected in Africa. Early giraffids migrated from Eurasia into Africa by way of what is now Ethiopia. Nine genera of fossil giraffids and five species of *Giraffa* lived in Africa, but all of these are now extinct except for okapi and our hero (Churcher, 1978; Mitchell and Skinner, 2003).

One of the earliest giraffids from the early Miocene (about 18 mya) in Libya was *Zarafa zelteni*, a lightly built animal rather like an antelope with ossicones sticking out on either

---

[2] Birds and reptiles usually evolved long necks by increasing the number of their neck vertebrae.

[3] From a hypothetical point of view, the tallest organism in the world which could breathe and move about would be 3.6 m (Page, 2012).

(a)

(b)

(c)

(d)

(e)

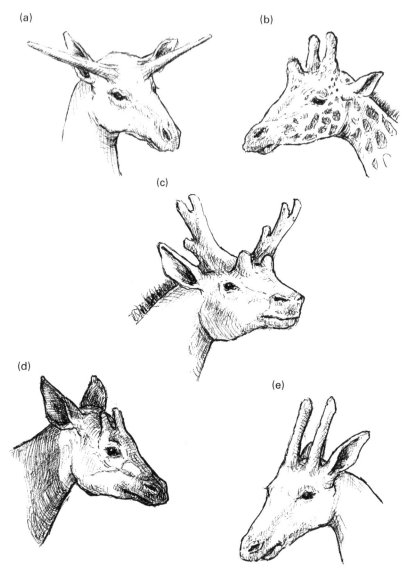

Figure 1.1   Giraffids from Africa: **a**, *Zarafa zelteni*; **b**, *Giraffa camelopardalis rothschildi*; **c**, *Sivatherium*; **d**, *Okapia johnstoni*; **e**, *Giraffa jumae*. After Churcher, Dagg and Foster, 1982; artist Jean Stevenson.

side of its head. Its line gave rise later to the okapi. By contrast, *Sivatherium maurusium* was the largest and most massive giraffid, as solid as an elephant and standing 2.2 m high. It was a more successful species because it lasted in Africa from the Pliocene to the late Pleistocene, about 8000 years ago; its fossil bones and teeth have been unearthed in South Africa, Uganda, Kenya, Ethiopia and Morocco, one cache spectacular because it involved 165 specimens (Franz-Odendaal *et al.*, 2004). This species had heavy spreading ossicones on its head, plus two smaller bumps in front of them, but it did not have a long neck (which would have been unable to support a weighty head) (Switek, 2009).

Giraffa jumae was apparently the earliest giraffe, existing from the Miocene to the mid Pleistocene, with fossils discovered in Tunisia, Ethiopia, Kenya, Tanzania and South Africa (Churcher, 1978). It was slightly larger than our present-day giraffe, with long legs but no median ossicone on the forehead. The earliest known bones of *Giraffa camelopardalis* are from the late Pliocene in Chad, from the early Pleistocene in Algeria and from the mid Pleistocene in Kenya and South Africa. Giraffe lived in Morocco until 600 AD, but then the drying of the Sahara made conditions impossible for it.

Formation of races

Recent research involving giraffe DNA tells the story of *Giraffa camelopardalis* as it spread out across the continent and evolved into different races (discussed in detail in Chapter 11). In their ground-breaking research, David Brown, Rick Brenneman and their colleagues (2007) of the International Giraffe Working Group analysed genetic material from 266 giraffe from 19 localities in Africa. This involved biopsy punches whereby a small amount of tissue was collected from each animal after it was darted in the flank or neck (Karesh, 2008). The researchers found that mitochondrial DNA sequences and microsatellite allele frequencies correlated with sharp geographic subdivisions. The divergence of these animals occurred in the mid to late Pleistocene when there was intense climate change in sub-Saharan Africa. Brown and his colleagues (2007) mention three specific climate-related factors that could have caused this diversification over time:

(a) in the late Pliocene conditions became cooler and drier which probably reduced connections between habitats frequented by giraffe;

(b) pronounced periodic oscillations (with a 21,000 year periodicity) of wet and dry conditions driven by changes in the intensity and location of maximal insolation may have facilitated habitat fragmentation and population isolation; and

(c) smaller-scale changes in habitat distribution perhaps promoted the isolation of specific populations, such as the expansion of a Mega Kalahari desert basin during dry periods of the late Pleistocene, which could have isolated the Angolan from the South African giraffe populations.

These climatic fluctuations, which presumably caused widespread changes in vegetation and habitat, leading to divergence, are evident in species other than giraffe, such as the hartebeest (*Alcelaphus* spp.).

For the past 50 years, no one doubted that giraffe all belonged to one species, *Giraffa camelopardalis*. Giraffe from specific regions of Africa often have similar spotting patterns (although never identical ones). As well, giraffe from all parts of Africa breed freely with each other in zoos. Giraffe have long legs that can carry them long distances; there seems to be no physical reason why, in the past, individuals from different parts of Kenya where there are now three races, for example, might not have interbred fairly often, given that there are no large rivers or mountains to prevent them from coming in contact.

Brown, Brenneman and the International Giraffe Working Group knocked this idea for a loop. From mitochondrial analysis, the team studied 381 individuals from 18 localities.

Figure 1.2   Rock-engraving of giraffe from Tuareg, Sahara. From Laufer (1928).

Only three of these animals were identified as hybrids between adjacent groups: one Masai/reticulated and two Rothschild's/reticulated animals. This emphasizes the low rate of interbreeding between subspecies/races. Indeed, even interbreeding among giraffe groups in the same ecosystem may be rare or non-existent judging from genetic material from members of four groups of Masai giraffe in the Serengeti that were only 60–130 km distant from each other. This has implications for research on the social behaviour of various races, as we shall see in Chapter 4.

The genetic data show that interbreeding among the current accepted subspecies of giraffe is almost unknown (0.8%). They rather are separately evolving lineages with

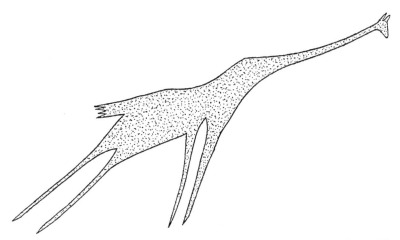

Figure 1.3   Rock-carving of running giraffe by Bushmen, South Africa. From Laufer (1928).

their own ranges. From mitochondrial-based coalescence analysis, it appears that these groups have been separated from one another from between 0.13 and 1.62 million years. Because of this, giraffe may be categorized in the future as being of more than one species rather than one single polytypic species. Indeed, the team suggests that the Masai giraffe might actually consist of two species, one on the east and one on the west side of the Rift Valley, and that some subspecies may have distinct population units within them. It is incredible to realize that this large mammal that spends its life constantly moving from one place to another while browsing actually covers a small area during its lifetime, and because of this has a taxonomy similar to that of a highly sedentary species.

Such a low rate of gene flow among different subspecies of giraffe living not that far apart is unprecedented for such a large, highly mobile African mammal. Why might this be? One suggestion from Brown and his colleagues is that reproductive cycles in giraffe (as in most large herbivores) are such that young are often born coincident with the growth of new browse rich in protein during the annual climate cycle. This occurs in July and August in the north, but in December–March in the south. The mortality rate of young giraffe is high (up to 67% a year: Leuthold and Leuthold, 1978a), so the faster they grow, the better. Should giraffe from north and south interbreed, young could be born with reduced fitness owing to environmental conditions. However, as we shall see in Chapter 9, giraffe can be born in any month; it may be that conception rather than birthing time tends to occur during the optimum growing season.

Another isolating factor may be the coat patterns of giraffe, remembering that communication among these animals is mainly visual. Calves who stay close to their mothers during their first few months may become imprinted on her spotting pattern and, of course, that of themselves. This could affect their choice of mating partners when they become adults. But again, the coat patterns of different races are often not that different.

Even food preferences might be important in separating giraffe populations. *Acacia tortilis* is a highly favoured food for Masai giraffe in the Serengeti (Pellew, 1983c), for example, whereas a population of *G. c. giraffa* browsed few of these trees in South Africa (Bond and Loffell, 2001). A map can be found in Chapter 11.

Such minimal interbreeding raises another interesting question. Until recently, many short-term studies of giraffe in the field have found that individual giraffe do not tend to hang around with other individuals to any extent. In his study of giraffe in Nairobi National Park, Bristol Foster documented that although giraffe were usually seen in groups, the members of these groups were constantly changing, which made him think that giraffe were not very sociable (Foster and Dagg, 1972). If two females were seen together in the Park more than a few times, these sightings were usually not successive, indicating that the two were far more often apart than together. One imagined them going off for days or weeks in different directions to browse, then wandering back to the Park as the spirit moved them. These new findings of minimal interbreeding between races, however, seem to indicate that giraffe are members of groups much larger than human researchers have been aware. It may be that giraffe breed within their very large groups rather than males, especially, going farther afield to find females in oestrus. This also implies that individuals recognize the scores or hundreds of other giraffe in their group, even though they do not see them particularly often. This indicates a fission/fusion system as is present in chimpanzees (Boesch, 2009), with extreme emphasis on fission rather than fusion.

Thinking of giraffe as belonging to many more species rather than one is far more important than just a matter of taxonomic interest. The number of total giraffe is decreasing each year because of pressure from expanding human populations and poaching, topics discussed in Chapter 11. Now we know that the decrease of some of these distinct populations is so great that they have become officially highly endangered – *peralta* with perhaps 300 members in Niger and Rothschild's population of about 700 animals (Brown, pers. comm., 2012). Every effort must be made to ensure they do not become extinct. It is no longer enough to think that there are many thousands of giraffe in national parks and reserves throughout Africa so what is the problem? Instead, we must ensure that all genetic populations remain viable.

Whereas nine subspecies have been accepted for many years (Lydekker, 1904; Dagg, 1962c; Dagg and Foster, 1982), now David Brown and his colleagues (2007) suggest the tentative number of possible species (but still considered subspecies at present) is six:

> *G. c. angolensis* from southwest Africa;
> *G. c. giraffa* from southeast Africa;
> *G. c. tippelskirchi*, the Masai giraffe from southern Kenya and northern Tanzania;
> *G. c. rothschildi* in Uganda and western Kenya;
> *G. c. reticulata*, the reticulated giraffe from northeast Africa; and
> *G. c. peralta* from Niger the northwest area of Africa.

Samples of their mitochondrial DNA sequences show that the first three southern subspecies and the last three more northern ones are more closely related to each other than to members of the other group. Giraffe of uncertain status include *G. c. antiquorum*, the central African form, animals from Zambia (*G. c. thornicrofti*) and those from Nubia and the Congo area.

Russell Seymour (2010), however, notes that the Masai giraffe, *tippelskirchi*, is actually more closely related genetically to the southern group despite its presence in Kenya and Tanzania. He also reports that the Zambian race, *thornicrofti*, is grouped within the southern clade but shares a mtDNA haplotype with giraffe from both north and south of Tanzania. Therefore, in the past, there must have been a migration of giraffe across the 'suboptimal habitat barrier'. Seymour also discusses tentatively the genetic relationships within the northern and southern groups.

The current status of and conservation measures for the races of giraffe are discussed in Chapter 11. Further information on recent genetic discoveries is reported by Liu *et al.* (1996), Vermeesch *et al.* (1996), Hassanin *et al.* (2007) and Huang *et al.* (2008).

Giraffe in European history

Giraffe were considered exotic and precious creatures to the early civilized nations of Egypt and Europe (Dagg and Foster, 1982; Mitchell, 2009). It was difficult to catch one in Nubia without hurting it to send down the Nile to Egypt, and demanding work to find it enough browse while in captivity to keep it healthy. Even so, individual giraffe were imported into Egypt as early as 2500 BC at a time where they no longer occurred there naturally. The two hieroglyphics representing the giraffe were translated as *ser* or *mimi*. The head of a giraffe on the body of a man was used in early times to represent Set, the god of evil who lived, like giraffe, in desert areas. That Set was always drawn with a long snout and square ears puzzled Egyptologists; what animal has square ears? It was finally decided that the ears were in fact horns, and therefore the animal must be a giraffe.

Queen Hatshepsut was given one for her zoological gardens in Alexandria by the region she conquered further south, and one was included in the Grand Procession (alias parade) in Alexandria about 270 BCE. Ostensibly it was given in honour of the god Dionysus by Ptolemy II, the Greek king who had recently taken over Egypt, but his real aim was to let his enemies know that this was a great and powerful new regime now that he was in command. It included 98 elephants marching four abreast, 48,000 foot soldiers, 23,000 cavalry, 2400 hounds led by slaves, 12 camels loaded with spice, 30 hartebeest, 24 oryx, 24 lion, 16 cheetahs and a single giraffe.

The first giraffe seen in Rome came from the zoo in Alexandria, given by Cleopatra to Julius Caesar in 46 BC. The animals were called 'cameleopards' in the advance publicity – they were as big as a camel but spotted like a leopard. Many Romans thought they would be as big as a camel but as fierce as a leopard, so they were disappointed in an age when the mass slaughter of gladiators and animals was valorized.

With the decline of the Roman empire, giraffe were forgotten in Europe. In 636 AD one scribe confused a cameleopard with a chameleon. Captive giraffe still existed in Egypt, but they were rare. When Egyptian forces defeated Nubians to the south, the annual tribute expected by Egypt was 360 human slaves and one giraffe.

Arab scholars wrote about this species, but few of them had ever seen a *xirapha*, the Arabic name of 'one who walks swiftly' for this creature. One Arabic pedant noted in 1022 that the father of a giraffe was a leopard and the mother a camel. He scorned those

Figure 1.4   Giraffe depicted from an Egyptian tomb. From Laufer (1928).

Figure 1.5   Giraffe with guide from Roman mural painting. From Laufer (1928).

who thought a horse could father a giraffe. 'It is well known that horses do not mate with camels any more than camels mate with cows.' Another Arab wrote that the offspring of a cross between a male hyena and a female camel had to be crossed in turn with a wild cow to produce a giraffe. It is clear that pedant Arabs had little to do with domestic stock.

Figure 1.6   Giraffe from Egypt depicted in 1554 by André Thevet. From Laufer (1928).

Persian scribes wrote other fantastic tales; one was that when a young giraffe was born it immediately fled from its mother so that she would not flay it with her rough tongue. It only returned to nurse when it was three or four days old.

Giraffe were reintroduced into Europe in 1215 when the Sultan of Egypt gave one to Frederick II, the Holy Roman emperor, in exchange for a polar bear. Frederick sent this or another giraffe on tour to other European cities and towns. From that time on, for nearly seven centuries, all giraffe to reach Europe were presents from Eastern potentates. One that reached Italy in the fifteenth century was given a test by Cosimo de Medici to see if it behaved more like a camel than a leopard. He put the giraffe, several lions, some dogs and some bulls together in a pen to see which species was the most fierce. Fortunately, the carnivores were not hungry so the giraffe survived, but it trembled with fear so was deemed to be most like a camel. Another giraffe was sent to Lorenzo the Magnificent, who had it stroll with its keeper through the streets of Florence, to the pleasure of the townspeople. Competition was fierce among the Italian zoos, so Florence was triumphant when it acquired a second giraffe because Naples, for example, had only one.

Several hundred years later, in 1597, a Scotsman visited a captive giraffe in Constantinople that was too familiar: 'He many times put his nose in my necke, when I thought myselfe furthest distant from him, which familiarity of his I liked not; and howsoever the keepers assured me he would not hurt me, yet I avoided these his familiar kisses as much as I could.' A London merchant wrote about this same animal that 'two Turkes, the keepers of him, would make him kneele, but not before any Christian for any money'. Another man from Flensburg painted a watercolour of the male, writing under it 'A strange and marvelous beast, the like of which we had never seen before.'

A giraffe was given by Bengal (part of India) to China in 1414 as one of a series of diplomatic relations between the two nations at that time (Church, 2004). In 1600, more were shipped to China where they were much appreciated, being considered as auspicious omens that promised good government, rich harvests and a peaceful reign. A Jesuit priest wrote: 'When any come to see them, they willingly and of their own accorde, turn themselves round as it were of purpose to shewe their soft haires and beautifull coulour, being as it were proud to ravish the eies of the beholders.'

Two centuries later, in 1805, apparently the first giraffe reached England, brought by an animal dealer who planned to become rich by displaying this wondrous beast. Sadly, it died after only three weeks, but in any case was said later only to be a white camel artificially spotted to look like a giraffe.

In 1826 the Pasha of Egypt sent two giraffe to Europe because so many European traders, officials and officers had admired them; he had the consuls of various countries draw lots for the animals, which Great Britain and France won. So a second (?) giraffe was shipped to London where it was housed at Windsor so King George IV could often visit it. The king was thrilled with this gift, often talking about it to the Duke of Wellington who was trying to discuss royal business matters. This animal too did not fare well, not because it was too cold or given the wrong food, but because when it had been captured in Africa it was sometimes roughly tied on the back of a camel. After two years it could no longer stand, so it had to be hoisted to its feet by ropes and pulleys for an audience with the king. When it died soon after it was stuffed with tow. This giraffe was painted by both Richard Barrett Davis and Jacques Laurent Agasse. In the Agasse painting, now owned by Queen Elizabeth II, it looks via Google to be in splendid shape despite its disability.

The giraffe that was won by the French in the lottery had a much fuller life than its friend who went to Britain. Her story has been told in many articles as well as a book (Dagg, 1963; Allin, 1998). She sailed from Alexandra to the quarantine station outside Marseilles accompanied by a French consul, four Arabs and three cows to supply her with her customary 20 litres of milk a day. After a short stay there, her home became a wooden pen in the city where she lived during the winter, going for a walk each fine day led by the Prefect of Marseilles who had designed a body cloth embroidered with the Arms of France for her to wear.

On 20 May 1827, she set off for Paris with her earlier retinue to which was added the zoologist Etienne Geoffrey-St-Hilaire, the Prefect, an African from Darfour, a mulatto interpreter, a Marseilles groom and police from the area through which the parade was passing. These were followed by a cart full of beans, barley and wheat to supplement the giraffe's diet of milk, and several other rare animals also walking to Paris. Jubilant farmers and townsfolk lined the route and crowded around the inns where the party spent the night. When they reached Paris on 30 June, troops had to hold back the populace which was agog to see this wondrous creature.

Parisians were mad for the giraffe. The rich began to wear gowns, necklaces and waistcoats in the colour of her coat, and spotted dresses, gloves and handbags to celebrate the new fashion '*à la girafe*'. When she exercised in the parks with her African attendants, tickets were issued to keep the crowds in control. In July, escorted by the professors at the Paris Museum, she walked to Saint-Cloud where the king was living. He was so delighted to see her that he allowed her to eat rose petals from his hand. The giraffe settled in at the Jardin des Plantes in Paris where she lived for 17 years. When she died, her body was stuffed with tow and given to the Verdun Museum. It was destroyed, along with the museum, during World War I.

In the early 1830s France had a fine giraffe, but England had none. Rivalry. So in 1836 King William IV commissioned a Frenchman, M. Thibaut, to go to Africa to collect some. Along with an Arab, three Nubian helpers and some local men with swords, the group managed to kill a female in the Kordofan Desert and capture her calf. It was fastened to the Arab for four days until it was tame enough to follow behind the caravan on its own. It was finally taught to suck milk from the fingers of the Arab so from then on it survived on camel's milk. Eventually the group caught three more calves. All four sailed down the Nile to Cairo and on to Malta where they were accommodated until the Lords of the Admiralty could arrange to ship them to London. Mr Thibaut was paid £750 for his efforts, a huge sum in those days.

Their arrival at the docks and march through London to the zoo in Regent's Park caused a sensation. Each giraffe was supposed to be led by two keepers holding long reins, but the giraffe walked too fast, their first exercise in weeks, so these men were unceremoniously dragged along after them. When they spotted the trees in the Park they became very excited, waving their necks about and kicking with their legs. It was all Thibaut could do to lure them into the elephant house with bits of sugar. By the following year a giraffe house had been built for them into which they moved. They and their progeny lived there for many years, the last descendant dying in 1881.

Giraffe were later sent for exhibition further afield as a way to make money for entrepreneurs, the first three reaching America from Cape Town in 1837 (Kisling, 2001). Between 1866 and 1886, Carl Hagenbeck Jr imported 150 giraffe to Europe, during which period Hamburg in Germany superseded London as the centre for international animal trade (Reichenbach, 1980). At the turn of the twentieth century Hagenbeck, who had become a millionaire thanks to his importation of exotic wildlife to northern countries, sent the first giraffe to Japan (Kisling, 2001).

Giraffe were new and exciting to people who had never seen them before, but not to those in the huge areas in Africa where they lived (Dagg and Foster, 1982). They were killed for food well before guns were introduced to the continent. Arabs hunted them on camels or horses, often in relays because giraffe can sprint fast and gallop for some distance at slower speeds. If a mounted Arab caught up to one, he could cripple it by cutting a hamstring muscle with his sword. In Ethiopia and Sudan, men surrounded a group of giraffe with fire, then chased them over a cliff where they fell to their death. Wakamba surrounded herds of giraffe, then killed a few as they broke free. Ndorobo caught giraffe in camouflaged pitfalls, ideal for tall animals that don't peruse their path as they stride along. Giraffe were killed by a poisoned arrow if an assailant could approach them closely enough. Many tribes use snares even today to kill wildlife; if the neck or a leg of a giraffe is caught, the wire will fester and poison the animal, dooming it to a slow and painful death.

A dead giraffe offered a wealth of opportunity – explorer Henry Stanley cut off 452 kg of fresh meat from one large male, although the flesh of females or young would have been more tender. Dried meat was preserved as biltong or ground into an edible powder. The long leg bones contained much marrow, with the bones themselves later used for fertilizer. Leg tendons were used for sewing, or as guitar or bow strings. Giraffe hides were made into pots, buckets, drum covers, whips, sandals, amulets and shields; shields made from giraffe hide were lighter than those from buffalo or rhino hides, but still tough enough to resist swords or lances.

The long tail of the giraffe was especially valued for switches and ornaments. Hairs were used as threads by Masai women to sew beads onto clothing, and other Africans made necklaces, bracelets and amulets. (I have zoo-giraffe hair bracelets made by friends, which bring me good luck.) In Chad, a sultan who owned many giraffe tails was known to be rich because they reflected the value of his horses that had pursued the giraffe.

Native Africans did not kill enough giraffe to jeopardize their populations. European men arriving with guns did infinitely more harm to the species, threatening the extermination of some races. Thousands were slaughtered for no good reason by explorers, settlers and white game hunters. White men in Kenya chased 20 giraffe into a swamp where they became so entrapped by the mud that the men could cut off their tails before leaving them to die of starvation (Johnson, 1928). Others cut off the neck and heads of giraffe they killed so they could stuff them to mount on their wall. Odd to be proud of having shot an animal as defenceless as a domestic cow.

Boer horsemen in the south were equally cruel in the name of sport, running down and shooting whole herds, often 300 or more dying in one year (Schillings, 1905). In South Africa there were many thousands of giraffe in 1800; a century later there were so few that Boers were forced to import expensive giraffe hides from East Africa to make their sjamboks (Schillings, 1905). One white hunter who claimed that the chase offered 'a sufficiency of sport to satisfy the most ambitious hunter' shot four giraffe in 15 minutes (Bryden, 1893). Another insisted that 'every European hankers after the killing of at least one if not several specimens of giraffe' (Schillings, 1905). In reality, the excitement of the chase seemed to involve mostly keeping one's clothing from being torn by thorns and worrying lest one's horse should trip during the pursuit.

Fortunately, just before it was too late, there was a change in sentiment and law toward wildlife. It was obvious that people had caused the extinction of the passenger pigeon and the near extermination of North American bison. African wildlife could have gone the same way easily, especially since a plague of rinderpest, brought in by cattle, swept through much of Africa in the 1880s and 1890s, killing vast numbers of animals. In 1933, the first Conference for the Protection of the Fauna and Flora of Africa was held in London. Soon countries were setting aside reserves and parks for their wildlife and punishing poachers more or less severely. Even today, however, giraffe can be legally shot by tourists in Zimbabwe, South Africa and Namibia.

Giraffe numbers increased in most countries during the 1900s because of protective measures. Tourism eventually became an important player in the realm of giraffe, with visitors from around the world bringing in money to see these and other animals in the wild. Some individual giraffe became buddies with people (Dagg and Foster, 1982).

Shorty, living in Hluhluwe Park in the 1950s, allowed children to put hats on his head and to sit on him as he lay down.[4] Another giraffe near the Blue Nile visited the biologist Brehm on his barge each day to be petted and given snacks. Betty and Jock Melville set up the garden around their large house near Nairobi as a refuge for Rothschild's giraffe which were becoming scarce in the wild (Patterson, 1977). Today the house is a small hotel surrounded by 47 ha of land where about 12 giraffe roam, thrilling visitors when they poke their head through a window to be fed a treat (Anon., 2012d).

Unfortunately, the positive measures for giraffe in the 1900s have been largely reversed in the present century as will be detailed in Chapter 11. Giraffe are not now routinely shot as they were in the past, but their numbers are devastated by poachers using snares or guns, farmers protecting their crops, settlers moving into their home ranges, deforestation, trains, disease, droughts, accidents and vehicles.

Giraffe are no longer perceived by the Western world as a good source of protein as they were in the past; giraffe numbers are declining so rapidly that huge efforts are now focussed on preserving all races of giraffe which might otherwise become extinct. Exceptions are Zimbabwe, Namibia and South Africa, where giraffe populations are not endangered, and where rich tourists now come to shoot them with bows or high-powered guns (Owen, 2012). Safari clubs and game reserves ask for a £1500 trophy fee, and by the time transportation, resort costs, guides, trackers and skinners are paid, the total cost for a hunting expedition can equal £10,000. It can be argued that in the few countries where giraffe numbers are actually increasing, the killing of males especially, which supply large quantities of meat, is good for the economy and does not put giraffe populations in jeopardy. Indeed, if game animals were subject to management and cropping on a commercial scale, they would greatly outstrip the meat provided by domestic cattle (Foster, 1968; see discussion and references in Dagg and Foster, 1982, pp. 24–25). The biomass of game animals in some areas of East Africa can be twice that where only domestic stock is run (Lamprey, 1964), and giraffe carcasses can supply a large volume of meat prepared in part as standard cuts (ugh) (Hall-Martin *et al.*, 1977a).

We must work hard to protect all giraffe populations and subspecies from extinction. This time, as never before, many scientists, government-sponsored agencies, local groups and international associations are cooperating to ensure the future of giraffe. These groups are listed at the end of Chapter 11.

---

[4] Shorty will always be close to my heart. It was only because a zoology professor at Rhodes University, Griff Ewer, asked about his history when visiting Hluhluwe Park that she was told he came from the eastern Transvaal (now Mpumalanga) in South Africa. He was born near a property whose manager, Mr Matthew, finally agreed to let me come in 1956 to study the giraffe on his citrus and cattle ranch (Dagg, 2006). I was fortunate to meet Shorty in 1957.

2   The giraffe's environment

Over millions of years, evolution and the environment shaped the lives of the giraffids that lived in Africa. However, giraffids also helped to shape their surroundings. This chapter deals with environmental changes caused by nature and giraffe which have a hierarchy of importance:

(a) large-scale natural changes that can affect a huge area, using the Serengeti ecosystem in Tanzania as an example;
(b) mid-scale changes caused by giraffe living in areas within this ecosystem;
(c) changes induced on single trees by the foraging of giraffe; and
(d) changes to the giraffe's environment caused by human management (interference).

Large-scale changes in the Serengeti ecosystem

Giraffe search out habitats of trees and bushes where they can obtain sufficient browse for their needs, yet have enough open space so they can make a quick getaway should lions attack. Such wooded savannah areas occur in many parts of Africa, perhaps none more famous for its giraffe than the huge (nearly 11,000 km^2) Serengeti ecosystem. The recent history of the Serengeti demonstrates how radically vegetation (and therefore food sources) can change. Robin Pellew (1983c), now a prominent citizen of London with an OBE, describes three major events that made this environment especially attractive to Masai giraffe in the late 1970s:

(a) elephant invasions into the Serengeti Park, upending trees and destroying bushes while foraging;
(b) the end of a rinderpest epidemic in the mid 1960s, so that wildlife could again live and reproduce rather than possibly die of disease; and
(c) unexpected rainfalls in the 'dry season' which were (are) always welcome.

Hungry elephants can be a scourge. During the late 1950s and early 1960s they entered the Park in large numbers, driven there by African pastoralists and farmers who wanted to farm their previous haunts. These behemoths proceeded to knock over and destroy thousands of trees to obtain food; in the decade from 1962 to 1972, 26–50% of the mature forest canopy was demolished by elephants in the northern areas (Norton-Griffiths, 1979). The attractive *Acacia tortilis* with its spreading mature canopy had been the dominant tree of the Seronera woodlands, comprising about 41% of the plant species and 58% of the woodland canopy cover. Because of elephant destruction, mature trees over 6 m tall were reduced from 48% to 3% of the total tree population between 1971 and 1978. By 1978, about 94% of the *A. tortilis* population was less than 3 m tall, so still potentially vulnerable to fire. Within a period of two decades, a high-canopy wood-land especially suitable for giraffe had reverted to regeneration thickets. Not that the giraffe went hungry, however, because they also favoured the regenerated bushes on which they browsed along with other herbivores, suppressing the plant growth in the process. In South Africa, elephant and giraffe, both megaherbivores, browse many of the same plant species, favouring *Acacia karoo* especially in the dry season (O'Kane *et al.*, 2011).

Rinderpest, a virus that decimated countless animals in East Africa and to which giraffe are moderately susceptible, was finally eradicated in the Serengeti ecosystem by 1964. From that time on, ruminant populations expanded dramatically, especially wilde-beest and buffalo. This increase correlated, after the mid 1960s, with a natural decline of wild fires, caused in large part by migrating wildebeest eating up to 85% of the low vegetation as they moved through the Serengeti. With reduced fire, shrub and tree regeneration flourished, to which ungulates including giraffe populations responded. However, by 1980 the development of tree regeneration was being checked by giraffe browsing at a height where it was still potentially vulnerable to fire.

Unseasonal rainfall was also a huge bonus for wildlife; between 1972 and 1978 rainfall patterns changed, with an increase of rain in the dry season and a decrease in the wet season, making the 'dry season' far less perilous for ungulates than it had been. This period was exceedingly prosperous for giraffe. Judging from aerial surveys, between 1971 and 1976 their rate of population increase was about 5–6% per year. By 1977, giraffe densities in the Serengeti National Park had reached a very high value of 1.19 giraffe per km^2 (Pellew, 1983b). Tree regeneration continued to flourish in the 1980s, so giraffe benefited from that too. In 1977, however, the border between Tanzania and Kenya was closed for political reasons. With many fewer tourists visiting the Serengeti there was less money to finance anti-poaching squads, so snaring and shooting of game increased (Sinclair and Arcese, 1995). Although illegal, the trade in bush meat remains serious, given the huge number of people living in poverty in the region and beyond. Even so, more recently giraffe numbers are said to have only declined 'a small amount since 1971' (Sinclair *et al.*, 2008, p. 503). However, in 2010 there were judged to be only about 5000 giraffe in the Serengeti ecosystem (Strauss, 2010).

Changes in habitat caused by giraffe

How does one measure biomass available for herbivores? While carrying out research for his doctorate from the University of Cambridge, Robin Pellew (1983a) studied in depth the food resources available for giraffe in part of the Serengeti National Park. He carried out extensive observations of feeding behaviour of giraffe, but he also studied the ways in which, by browsing, giraffe change their habitat. For his research area of 120 km^2 *Acacia*

Figure 2.1   Terrain of Rothschild's giraffe with yellow fever tree (*Acacia xanthophloea*). Photo by Zoe Muller.

woodlands, he wanted to know what plants were present, the biomass of these species in wet and dry seasons, which species were browsed by giraffe and how their browsing affected the vegetation itself.

Pellew (1983a) documented the potential food available to Masai giraffe in four main feeding areas identified from their drainage patterns:

(a) ridge-top and upper-slope *Acacia* regeneration thickets;
(b) mid-slope open *Acacia/Commiphora* woodland;
(c) valley-bottom *Acacia xanthophloea* riverine woodland along main rivers (the Upper Seronera River and its tributaries); and
(d) valley-bottom *Acacia* regeneration thickets along secondary watercourses.

Some trees favoured by giraffe for food were fenced off in three 25 × 25 m enclosures to prevent their browsing; thus any new growth could be measured every 3 months (collected from the central 20 m × 20 m of the enclosures – giraffe have a long reach) to help estimate browse production. These highly favoured giraffe food species were *Acacia tortilis*, *A. senegal*, *A. hockii*, *A. xanthophloea*, *Grewia bicolor* and *G. fallax*.

Vegetation biomass depends on sunlight and rainfall, the latter reaching a peak during the wet season. Plant growth decreases rapidly in the early dry season because of the secondary thickening of shoots produced in the wet season and the very low level of new shoot production. The biomass density is greatest at the top canopy height where light is most intense; however, about 40% of the biomass available to giraffe fell below 2 m and was thus potentially available to smaller herbivores too.

Pellew measured the browse production in two ways: (a) the mean annual shoot production with no clipping of shoots (unbrowsed production) and (b) the mean annual shoot production with shoots clipped at quarterly intervals to simulate giraffe browsing (clipped production). By comparing these two sets of data for one year's growth, he found that clipping stimulated shoot production for *Acacia xanthophloea* by 48%, for *A. tortilis* by 70% and for *A. hockii* by 81%. Cattle saliva contains plant growth-promoting agents, and the copious quantities of viscous saliva which giraffe leave on the cut surface of a shoot may do this too. Pellew estimates that in his study the riverine woodland produced each year about 5000 kg per ha of browse available to giraffe, while the ridge-top woodland produced only about 1750 kg per ha; the former value is significantly higher than other reported values for savannah browse. In the Serengeti at that time the overall effect of browsing by the giraffe was to stimulate the production of more forage available for herbivores.

How giraffe affect individual trees

When giraffe enjoy a hearty meal of up to 34 kg of browse a day (Muller, 2011a), they won't know that these apparently courteous trees and bushes have nasty ways to curtail this assault. Indeed, heavily browsed plants have evolved four means to accomplish this:

(a) the very shape of a tree may reflect the foraging of giraffe;
(b) the sharp spines or hooks on many plants deter giraffe somewhat from feeding on them;
(c) on some plants aggressive ants attack invading herbivores; and
(d) some trees browsed by giraffe produce toxic chemicals in their leaves.

Tree shape

Consider the tall flat-topped acacia trees such as *A. tortilis* that are characteristic of wooded savannah areas of East Africa and a favoured food plant of giraffe (Dagg and Foster, 1982). Only giraffe can take advantage of the foliage of trees over 3 m tall, and these will be confined to eating the leaves at the outer edges. If this flat-topped shape is indeed fashioned thanks to the browsing of giraffe, as seems likely, it has also given the trees and the giraffe other advantages. The wide crown maximizes the amount of sunlight energy available to the tree for photosynthesis. The crown itself, which casts a large shadow on the ground, helps the soil retain moisture as well as attracting animals to its shade during hot weather where they may drop manure. Various ungulates will graze favoured grasses there such as *Panicum maximum*, thereby reducing the fuel for fire and fire damage which can destroy young trees (Midgley *et al.*, 2001).

Figure 2.2   Savannah woodlands. Photo by Zoe Muller.

Spines and hooks

Many of the common trees of the open woodlands of Africa on which giraffe preferentially forage, such as various *Acacia*, *Balanites* and *Scutia* species, are armoured with spines or hooks. Evidence indicates that these weapons evolved because of heavy browsing by large mammals. To test this possibility, A. V. Milewski, Truman Young and Derek Madden (1991) on the Athi Plains southeast of Nairobi chose pairs of similar branches on *Acacia seyal* trees, on one of which they removed all the thorns. They found that hungry giraffe greatly favoured the branches without thorns. As well, for both *A. seyal* and *A. xanthophloea* trees in the area, thorns were longer at branch heights up to 4 m but significantly shorter at above 5 m, which giraffe cannot reach. Leaves were also longer at the higher height and at all heights near the research camp where giraffe did not venture.

Later, A. D. Zinn, D. Ward and K. Kirkman (2007) focussed on the natural browsing of about 100 giraffe on a game ranch in KwaZulu-Natal, South Africa, where they placed fixed-point cameras so they could determine the relative popularity of giraffe foraging sites. From these photographs they documented which areas had favoured *Acacia sieberiana* var. *woodii* leaves that were little browsed, partly browsed or highly browsed. The higher-browsed sites were in relatively open areas with plenty of water. The low-intensity browsing sites occurred where the vegetation was very dense and there was no water source. In addition, two sets of samples of leaves and spines growing above a height of 2.5 m (which only giraffe could reach) were taken from 20 randomly chosen acacia trees. The spines were then measured along with the chemical composition of the leaves. The results showed that the trees with high browsing intensity had significantly longer spines and smaller leaves. The tasty trees had shown an 'induced defence' reaction because (a) their response was associated with herbivory, (b) their response reduced herbivory, and (c) reduced herbivory increased the plants' fitness.

Stinging ants

The acacia trees beloved of giraffe are often associated with stinging ants (*Crematogaster* spp.), an example of a symbiotic relationship between ants and plants that has been well documented over many years (Beattie, 1985). Some acacias have extra-floral nectaries in the form of a small pad at the base of each petiole which provides room and board for ants. Other ants live in thorns or in 'galls'. Madden and Young (1992) studied *Acacia sieberiana* trees on which giraffe in Kenya like to browse. They found that although ants swarmed onto the head of animals chewing leaves near their base, they did not bother subadult or adult giraffe who kept on browsing, but they did drive away young giraffe. A calf might suddenly stop feeding on an *Acacia drepanolobium* tree and leap back, violently snorting and shaking its head.

Madden and Young also looked into the question of biting ants in the semi-arid game ranch southeast of Nairobi where two main species of acacia trees attracted browsing by giraffe. Unlike *A. seyal*, *A. drepanolobium* trees had some thorns with swellings that were often inhabited by ants. Both trees had leaves of about the same length, averaging 2.5 cm, but *A. drepanolobium* had thorns much shorter than the leaves, while *A. seyal* had thorns as long or longer than its leaves. Their thesis was that *A. seyal* had long spines to deter feeding by herbivores, while *A. drepanolobium* had less need of such thorns because it had ants to carry out the same mission.

In mornings or late afternoons when the free-ranging giraffe were hungry, the researchers observed individuals feeding at *A. drepanolobium* trees, noting when each touched a leafy branch with its muzzle to begin browsing and when it stepped away again. Then they shook the branch many times with a smooth stick employing the same force that the giraffe had used in order to collect the aggressive ants that crawled onto the stick. The ants *Crematogaster mimosae* and *C. nigriceps* (?) emit a distinctive odour when disturbed as well as bite, wiping their sting in the wound to produce irritation (noted when angry ants crawled under the researchers' clothing). Some *Tetraponera penzigi* ants also occurred in the stem tips of some of the trees, but they were reluctant to bite or leave the foliage. The researchers concluded that although the larger giraffe didn't seem deterred by the ants, they may indeed have been to some extent, so further research is needed.

Chemical defence – tannin and poisons

Giraffe and other African browsers like to feed on species of acacia because they have a high crude protein content. However, as a response to heavy browsing, many acacia trees

Figure 2.3   Arid savannah of Damaraland in Namibia. Photo by Julian Fennessy.

produce chemicals such as condensed tannins, cyanogenic glycosides and, in some species, prussic acid to fend off their attackers (Zinn *et al.*, 2007). Plants that have tannins in their leaves are usually a poor diet choice for browsers because the tannin binds to protein molecules, thus reducing nutrient availability after the leaves are ingested. Leaf samples of a tree that has been heavily browsed have high levels of condensed tannins in their tissues, indicating that this is a fixed defence of the tree. When coupled with leaves with low nutrient concentrations, tannins can provide an effective anti-herbivore mechanism. A little tannin can be good for ruminant diets because it enhances the function of the rumen and reduces internal parasite loads, but only limited amounts of tannins should be consumed (Brenneman *et al.*, 2009a). This defence of trees from over-browsing has been documented in Kruger National Park, South Africa, where plants favoured by giraffe produced tannins in their tissues which their consumers dislike in high condensed levels (Furstenburg and van Hoven, 1994). In Niger, nursing females specifically avoid eating leaves rich in tannin (Caister *et al.*, 2003).

Other chemicals negative to browsers that may be produced in leaves are cyanogenic glycosides; these can liberate hydrogen cyanide which is toxic to cell metabolism (Zinn *et al.*, 2007). Trees or bushes which had their foliage highly browsed had a higher total cyanide concentration in their leaves than did trees that were seldom browsed, so the trees or bushes had again in effect fought back chemically as a response to the trauma of being eaten.

Human interference/management

Human managers in Serengeti National Park wanted to increase the number of *Acacia tortilis* trees in the area because of their wide flat crowns that taste good to giraffe and look attractive to tourists (Pellew, 1983c). However, giraffe act against the replacement of mature *A. tortilis* trees because they delay its regeneration. A seedling of *A. tortilis* susceptible to browsing takes 36 years to grow beyond the reach of an adult bull giraffe (5.5 m), about 2.8 times longer than if it had not been browsed. To reach a height of 3 m, the maximum height of fire mortality, the seedling requires 21 years, compared with 8 years in the absence of giraffe.

Pellew (1983c) presents a model whereby the woodland structure desired by Park management of a mature canopy largely of *A. tortilis* can perhaps (given the many variables possible) be attained. The model results show that measures to promote plant regeneration by preventing fire and/or giraffe culling would be more effective in the long term in encouraging mature canopy recovery than would culling elephants in order to reduce the destruction of mature trees. The combination of low elephant densities and high giraffe densities prevalent in the Serengeti produces a dynamic system in which the woodland structure oscillates between mature canopy and open regeneration-grassland phases.

More recently, another management model has been proposed for the Sweetwaters Game Reserve in Kenya, which was formed and enclosed in 1988 to protect 25 endangered black rhinos (Birkett, 2002). By about 2000 the number of rhinos had risen to 90, which was wonderful, but so too had the numbers of other large mammals increased; there were now 150 reticulated giraffe and 100 elephants. The model predicts how many giraffe and other mammals should be translocated to preserve the natural vegetation for the rhinos.

Where giraffe populations are unnaturally large, as in some managed tourist areas where predators are few or absent, browse trees preferred by giraffe such as *A. davyi* may be completely wiped out, as happened in the Ithala Game Reserve in KwaZulu-Natal, South Africa (Bond and Loffell, 2001). There, 180–200 giraffe (presumably *G. c. giraffa*) roamed an area of over 29,000 ha of which roughly two-thirds was too steep or otherwise inaccessible to them, making a high stocking rate of about 1.8 giraffe per km². Compared to plant species that were thriving within no-giraffe areas nearby, in addition to no more *A. davyi*, a census of tree species available to giraffe revealed that most *A. caffra* trees within giraffe browse height were dead and *A. karroo*, the most common species, was heavily browsed. On this game reserve the browsing of giraffe was drastically altering both the composition of plant species and their distribution.

Julian Fennessy (2011) notes that giraffe have also impacted their favourite forage source in Namibia, *Faidherbia albida*, causing structural changes in individual plants and in their distribution. This impact, along with damage by elephants and seasonal floods, has resulted in a shift in the age structure of the species over two decades.

This chapter discussed how giraffe shape their environment by the choice of plant leaves they browse. The next chapter considers the environment from the point of view of giraffe: not what they do to the environment, but what the environment can do for them to make them grow and thrive.

3 Feeding in the wild

Foraging

Feeding is necessarily the prime activity of giraffe in the wild because they have a large body to maintain. They may eat at every hour of the day, especially in winter or dry seasons when less food is available, and in the night as well (Innis, 1958; Foster, 1966). Often they forage on leaves, flowers, fruits or twigs in the morning and late afternoon, spending the midday ruminating and resting. Although they are particularly fond of *Acacia* trees or bushes, they forage on many other species too, depending on the season. They do not select, as lesser kudu and gerenuk do, chiefly young and tender shoots (Leuthold, 1971; Leuthold and Leuthold, 1972). Barbara and Walter Leuthold (1978b), who over 16 days timed the activities of giraffe in Tsavo East National Park, Kenya, report that on average the eight males foraged for 27% of each day and the four females (most of them lactating) for 53%.

Giraffe have feeding down to a science. To consume a secondary branch with small leaves, say one of *Securinega virosa*, a giraffe curls its tongue around the base of the branch and, with a swift jerk of its head upwards and sideways, strips off all the leaves into its mouth, leaving the branch bare (Berry, 1973). If a branch has thorns, it nimbly works around them with its tongue to obtain the leaves, perhaps cutting a twig off with its broad lower canine. They eat bark too – along the Chobe River in Botswana they were stripping the bark from Natal mahogany trees (Tutchings, 2012).

If a fruit is very high, a male may reach up, trying to curl his tongue around the fruit's lower end before managing a grip that allows him to tear it free; or a giraffe may rear up on its hind legs or break thin branches to get at desired leaf morsels. Philip Berry (2007) observed this latter method four times by three adult *thornicrofti* giraffe, both male and female. In each case the leaves were too high for the giraffe to reach with its long extended tongue, so it grasped and bent or broke the lower bare part of the branch on which the leaves grew so that it could then strip the leaves into its mouth. On a fourth occasion, a younger male wanted to eat leaves from a vine but was obstructed from doing so by two other branches. He snapped each in turn by taking it into his mouth so he could reach in and extract the vine leaves he was after. Berry writes, 'this animal revealed remarkable cognizance of the best way in which to achieve his objective'.

Giraffe, like other wild ungulates, are extremely well adapted to their environment. For the Tarangire Game Reserve in northern Tanzania, Hugh Lamprey (1963), of the country's Game Division, reported 40 plant species eaten by Masai giraffe: 18 trees

Table 3.1 Plant species eaten by giraffe races in specific areas.

Reference	Location	Number of species if specified	Race
Lamprey, 1963	Tarangire Game Res, Tanzania	40	*tippelskirchi*
Foster, 1966	Nairobi National Park, Kenya	34	*tippelskirchi*
Leuthold and Leuthold, 1972	Tsavo National Park, Kenya	66	*tippelskirchi*
Wyatt, 1969	Nairobi National Park, Kenya		*tippelskirchi*
Pratt and Anderson, 1982	Arusha National Park, Tanzania	16+	*tippelskirchi*
Pellew, 1984b	Serengeti National Park, Tanzania	45	*tippelskirchi*
Young and Isbell, 1991	Athi Plains ranch, Kenya		*tippelskirchi*
Nesbit Evans, 1970	Soy, Kenya	21	*rothschildi*
Nesbit Evans, 1970	Maralal, Kenya	42	*rothschildi*
Verschuren, 1958	Garamba National Park, Zaire		*antiquorum*
Backhaus, 1961	Zaire	17	*antiquorum*
Ngog Nje, 1984	Parc National de Waza, Cameroon	22	*peralta*
Ciofolo and Le Pendu, 2002	Niger	45	*peralta*
Caister *et al.*, 2003	Niger	33	*peralta*
Klasen, 1963	Luangwa Valley, Zambia		*thornicrofti*
Berry, 1973	Luangwa Valley, Zambia	19	*thornicrofti*
Innis, 1958	Eastern Transvaal, South Afr	32	*giraffa*
van der Schijff, 1959	Kruger National Park, South Afr	43	*giraffa*
Brynard and Pienaar, 1960	Kruger National Park, South Afr	43	*giraffa*
Pienaar, 1963	Kruger National Park, South Afr		*giraffa*
Wilson, 1969	Matopos National Park, Zimbabwe	20	*giraffa*
Hall-Martin, 1974a	Timbavati, South Africa		*giraffa*
Kok and Opperman, 1980	Willem Pretorius Game Reserve, South Afr	25	*giraffa*
Parker *et al.*, 2003	Eastern Cape Prov, South Afr – extralimital	14	*giraffa*
Parker and Bernard, 2005	Eastern Cape Prov, South Afr –extralimital	49	*giraffa*
Cornelius *et al.*, 2012	Southern Cape, South Afr –extralimital	20	*giraffa*
Marais *et al.*, 2011	Southern Cape, South Afr –extralimital	22	*giraffa*

(including 8 *Acacia* spp. and 4 *Commiphora* spp.), 13 shrubs (including 3 *Grewia* spp.), 5 herbs and 4 grasses. There were 14 species of ungulates living in high densities in this reserve where sometimes food and water supplies were low, but there was little or no evidence of overstocking. Each species had different food preferences which caused ecological separation. This contrasted with the neighbouring cattle areas of the Masai Steppe where, because there was no such diversity, the habitat quickly deteriorated from the presence of too many animals.

For researchers, what giraffe eat in the wild is both easy and complicated to study. Easy because one can lounge in a vehicle near giraffe recording every morsel of plant material taken into their mouths. Many researchers have done this, sometimes also commenting on species preferred. Early lists tended to be short and tentative, but later ones named dozens of species browsed, which usually included acacias. These legumes have low levels of condensed tannin in their leaves and are rich in protein with a high water content, especially if they have new shoots.

Giraffe foods are complex to research because different areas have different vegetation. Some plant species are browsed by giraffe in some regions but not in others, and giraffe may choose to eat in areas for reasons other than the food that is there. They constantly move from one place to another, sometimes spreading out to browse during the rainy season when food is plentiful or, alternatively, spreading out during the dry season when it is not. Other imponderables include not only the availability of certain plants in an area, but such things as their stage of development, their taste and smell, their chemical composition, the familiarity of the food and an animal's physiology (Ngog Nje, 1984).

To determine plants browsed by giraffe in three nature reserves in South Africa, J. J. C. Sauer, J. D. Skinner and A. W. H. Neitz (1982) and Sauer (1983a) collected at 6-weekly intervals samples of vegetation eaten by members of the *giraffa* race. These samples, clipped from the same trees at about the same time of day for comparative purposes, were then analysed for moisture content, crude protein, ether extract, crude fibre, ash and nitrogen-free extract. Seasonal chemical changes in 15 plant species that giraffe ate at different venues were then studied further, because these were apparently favoured.

The researchers conclude that crude protein and a high moisture content were the two most important components that guided giraffe choice. They also note other insights:

- the moisture content of leaves from the same tree may have higher values at night than in the daytime, so that giraffe browsing nocturnally obtain more water;
- plant choice does not depend only on the chemical composition of the vegetation;
- *Acacia* and *Combretum* spp. are preferred by giraffe but are often unavailable because, being deciduous, they shed their leaves during the latter half of the dry season. (Sauer (1983b) notes that as *Acacia* leaves have the higher protein content, they are the better food source);
- the moisture content of most leaves decreased from December to about September and then increased again;
- with the decrease in moisture there was usually an increase in crude fibre and nitrogen-free extract in leaves; and
- occasionally, the chemical composition of vegetation sampled from the same species of tree at the same time of day varied radically when they came from different nature reserves (which was highly worrisome for the researchers, needless to say).

The researchers suggest further studies such as determining which leaves ferment most easily and fastest in the rumen, and which plant species produce the best ultimate spectrum of volatile fatty acids (the end products of fermentation), which are an important energy source.

British biologist Robin Pellew (1984b) carried out further nutritional studies in Tanzania. He found large differences in the nutritional qualities of food items available to giraffe and noted that these vary between seasons, between species and between plant parts of the same species. Browsing giraffe presumably continually adjust their feeding regimes taking into account these changes and also such things as leaf size, presence of thorns, time devoted to browsing and rate at which they chew.

Figure 3.1   Rothschild's giraffe feeding on high thorny branches. Photo by Zoe Muller.

In the Serengeti, free-ranging Masai giraffe fed on trees or shrubs for 96% of all feeding records, with herbs and vines contributing the rest (Pellew, 1984b). The females consumed twice as many herbs as did the males, while neither ate grass. In the wet season both sexes favoured *Acacia tortilis*, while in the dry season they switched to species of *Grewia* which grew low down. In all, the females had a nutritionally richer diet than did the males, while the bulls ate more high fibre and lignin content.

The giraffe did not select food items in proportion to their availability, but searched them out specifically – the six plant species they liked most altogether comprised only 3.6% of the total biomass present. Giraffe fed into the night, but to a lesser extent when there was no moon. They ruminated then at length, often lying down when doing so, and for brief periods sleeping with their head resting on their flank. Pellew includes in his paper the chemical composition of plants eaten and not eaten, and giraffe activities. They fed more in the dry season when browse selection was limited and spread out less during the wet season which gave them more time to relax and lie down. They reduced their movements during the hottest part of each day to conserve energy.

Pellew (1984a) calculated the energy budgets of giraffe feeding on available browse – not only what species they fed on, but how much energy this food provided for them. The daily rate of food consumption was similar to that of other ruminants – 1.6% and 2.1% of the live weights of adult male and female giraffe, respectively. The quality of their diet, assessed in terms of crude protein levels, was consistently higher than that of grazing ungulates, especially in the dry season when the protein levels of a browse diet showed

Figure 3.2   Giraffe also commonly forage on low vegetation. Photo by Zoe Muller.

only a marginal decline. This high energy budget explains the ability of female giraffe to produce young even in the dry season, when grazing species fall below the metabolic threshold for year-round breeding. Most herbivores in Africa are grazers, or both grazers and browsers, but because grass is less nutritious than the leaves of bushes and trees, they reproduce only at a set season each year when the most nutritious food is available for the female and her young.

Flowers as food

Giraffe often consume flowers when they are available. In the Kruger National Park at the end of the dry season, several acacia and other trees bloom so prolifically that the white flowers form a magnificent display, serving as a substantial source of food for giraffe (du Toit, 1990a). At peak season a quarter of their feeding time may be devoted to flowers of *Acacia nigrescens*; these flowers contain almost three times as much condensed tannin as do the leaves (a negative), but they also contain twice the protein content (a compensating positive). While browsing, giraffe collect pollen on their heads and necks which they may disseminate to other trees during subsequent feeding. Thus in theory the benefit to the plants of this pollen transfer could more than offset the number of flowers the giraffe consume.

Johan du Toit helped to arrange follow-up research to determine if giraffe were pollinators or flower predators of *A. nigrescens* trees in the Kruger National Park

Figure 3.3   Rothschild's giraffe feeding on a branch it has broken loose. Photo by Zoe Muller.

Figure 3.4   Namibian giraffe feeding near the ground with bent forelegs. Photo by Zoe Muller.

(Fleming *et al.*, 2006). To do so, the study group examined these flowering trees to determine the effect on them of giraffe browsing. Most of the flowering clusters were reachable and eaten by giraffe, but some trees were so tall that even the males could not catch hold of the blooms. These very tall trees had more subsequent fruit set than did the smaller trees with blossoms consumed by giraffe; browsing therefore had a detrimental effect on the fecundity of the trees with flowers within giraffe reach. Giraffe are indeed predators, although this word is unusual in this context.

Grazing

Giraffe rarely eat grass. It has seldom been observed, it is awkward for the animals to splay their legs to reach down near the ground and they are vulnerable to predation while in this position. However, Peter Seeber, Honestly Ndlovu, Patrick Duncan and André Ganswindt (2012b) based in Hwange National Park, Zimbabwe, found grazing to be a recurrent event in this nutrient-poor environment. They observed it for giraffe on 31 occasions, usually by females and by groups of 4–16 individuals. The animals frequently lifted their heads from grazing to look around in vigilance behaviour. Actual grazing time for each bout in the crouched position was 5 min (range 1–18 min) and the entire grazing time averaged 23 min (range 4–49 min) before the animals switched to another activity. Strangely, even while they were grazing there were nutritious leaves available nearby that they could have browsed. The authors postulate that perhaps grass contains

Figure 3.5   Dark Rothschild's female with calf under her belly eating prickly pear. Photo by Zoe Muller.

micronutrients and minerals unavailable in browse. (Curiously, grazing while lying down became a fad of giraffe in a Florida facility although it was counterproductive because they picked up parasites while doing so; see Chapter 10).

Pica behaviour

Pica behaviour, the eating of substances not considered food, is common for some giraffe which may lick or bite salty soil (geophagia) or chew on bones (osteophagia) (Western, 1971; Wyatt, 1971). Over a two-year period of watching giraffe near the Kruger National Park in South Africa, Vaughan Langman (1978) saw 60 of them lean down to search the ground before licking the earth or picking up bones or dirt with their mouths, then standing with their noses elevated to chew the material with their cheek teeth. Adults and subadults ate bones at about the same rate, but subadults were more likely than adults to eat soil. The warden in Nairobi National Park who must have observed pica behaviour made a point of spreading an unrefined mixture of salts from Lake Magadi so that game animals could partake of soda ash, sodium chloride, bicarbonate of soda and sodium fluoride (Dagg and Foster, 1982).

Because 90% of his sightings occurred from April to October, during the dry season when nutritious forage was more difficult to find, Langman suspected that pica was correlated with a dietary imbalance of calcium and phosphorus; calcium is readily available in many forms in plants and leaves but phosphorus may not be, although it is present in soil. There may be a correlation between pica behaviour and kidney stones, because of 75 giraffe from the area autopsied (which had died of natural causes or been culled), 29% had kidney stones and associated lesions indicating that phosphorus was likely deficient seasonally in the forage.

Closed ecosystems

In the future, as the human population of Africa increases, giraffe will have less room in which to live. Will the more limited food supplies be adequate? Many will find themselves in a smallish park or reserve which they cannot leave without danger of being killed. Rick Brenneman and colleagues (2009a) analysed this problem for Rothschild's giraffe in the Lake Nakuru National Park (LNNP) in Kenya. There had been 153 giraffe there in 1995, but by 2002 there were only 62, with many fewer calves than expected.

What had caused this decline? The Kenya Wildlife Service and experts looked into three possible causes beyond fires which were sometimes set by people.

(a) Inbreeding may have been involved because all the animals had been transported to the area from the same small population at the Soi army base. To check this out, skin biopsies of 18 giraffe were collected with biopsy darts for analysis. They indicated that there was no cause for concern because of inbreeding: the group was genetically healthy.

(b) Perhaps there was not enough suitable food for the giraffe? This seemed probable because there was over-browsing by giraffe of a preferred tree species, *Acacia xanthophloea*, both of its leaves and its bark. This likely began between 1993 and 1997, when there had been extreme heat and drought in the Park which at that time held the recommended carrying capacity of about 150 giraffe. Because of the climatic change, food would have been limited and food plants often over-browsed by the giraffe. Many acacias are known to produce increased toxic tannin in their tissues when they are stressed (Furstenburg and van Hoven, 1994). When lactating females fed on these acacias, the tannin content of their milk would have increased, to the detriment of their young.

(c) There were many lions in the Park, which perhaps brought down more young giraffe than usual. Calves would have been especially vulnerable had their health been compromised by a poor diet, which seems to have been the case.

The researchers note that under natural free-ranging conditions, giraffe during a drought will migrate to other areas in search of browse options. This was impossible for the Rothschild's giraffe, however, because the Park was fenced. They recommend that in an enclosed area forage availability and herbivore numbers be constantly checked, so that the recommended carrying capacity can be based accurately on these parameters. If drought conditions occur in the future in LNNP, they suggest, because the race is endangered, that the giraffe should be transported to another protected area with a surplus of browse. This would also enable the *A. xanthophloea* and alternate giraffe forage species in LNNP to recover naturally.

Food sources are particularly at risk in an enclosed area because browsing pressure is continuous. Katherine Forsythe and Laurie Marker (2011) studied the diets in north-central Namibia of 33 giraffe who lived in a 3650-ha fenced nature reserve. These animals did not feed at random on the available plants, but rather chose some species more often than others. Only 27% of the vegetation present was considered favoured, while 15–30% of their diet came from species they weren't too keen on. Gradually, because of selective browsing, the food choices of the giraffe would change the composition of the vegetation growing in the reserve.

Extralimital locales – stocking limits

Extralimital areas have closed ecosystems for giraffe which are of special interest because this species has apparently never before lived in these locales. Anxious to attract tourists, many private game reserves in southern South Africa have introduced giraffe to their properties. However, this new initiative raised questions. Would the giraffe thrive? How would they react to the available food, especially in winter when many species have lost their leaves? Would browsing by giraffe affect the composition of the vegetation, possibly destroying endangered species? After all, the vegetation had apparently evolved during a time when there were no large browsers such as giraffe. Perhaps most importantly, property owners needed to know how many animals their facility could support without degrading the vegetation.

Francois Deacon, T. C. van Wyk and G. N. Smit (2012) addressed this question in their research on the grassland biome of a wildlife property in the central Free State of South Africa. Over an 18-month period, they (a) estimated the browsing capacity of eight giraffe on the woody plants and (b) calculated the production of leaves and shoots of the bushes and trees to a maximum browsing height of 5 m. Using the formula of 1 BU (Browse Unit) = the metabolic equivalent of a kudu with a body mass of 140 kg browsing for one year on a hectare of land (Bothman *et al.*, 2004), they calculated that the estate had a 7.7 ha BU^{-1} for September, the month with the least browse availability. At the time it was estimated that all the browsers on the estate represented 70 BU, with 42 BU of these going to the giraffe, and the rest to kudu, nyala and impala. The property had too many animals on it. The effect of the overstocking was clearly evident, with broken branches and heavily browsed leaves and shoots. Obviously the vegetation would be even more degraded over time unless the number of browsers was significantly reduced.

Similar recent ecological studies have been carried out in extralimital facilities further south. Daniel Parker and several colleagues, R. T. F Bernard and S. A. Colvin (2003, 2005), based at Rhodes University in South Africa, focussed on private game reserves in the eastern Cape Province. Their research involved observers at three different translocation sites recording all the plant species on which giraffe browsed during three times of day in autumn, winter and spring. They noted 48 species consumed by giraffe, by far the most preferred being *Acacia karroo* (43%) and, because this plant is not so available in winter when it loses most of it leaves, *Rhus longispina* (17%), which is evergreen in all seasons. Do giraffe eat such a diverse diet, especially in winter, because they have little time to be selective? They certainly need a huge daily quantity of food to sustain their metabolic and reproductive requirements.

The researchers note that because of their browsing, giraffe did change the composition of the local vegetation. They also ate the leaves of one threatened species, *Dombeya rotundifolia* (such rare plants could perhaps be protected in the future by fencing). Because of these findings, in his Master's thesis, Parker (2004) recommends that giraffe numbers be limited in smaller properties in the Eastern Cape and that the combined effects on flora and fauna of all extralimital herbivores be continually documented.

In a mosaic thicket area of the southern Cape, South Africa, near Mossel Bay, Andri Cornelius, Laurence Watson and Anton Schmidt (2012) over a period of a year studied the diet of eight adults of the *giraffa* race who had been moved to Nyaru, a 429-ha private wildlife ranch. Seasons differ markedly this far south, so the diet of giraffe varies too (rather than varying with wet and dry periods farther north). Of the 20 plant species eaten, leaves of the deciduous shrub *Acacia karroo* were greatly favoured in the warmer months, while those of *A. cyclops* (an invasive evergreen alien) were eaten most in the winter. The researchers suggest that the stocking rate of giraffe could be related to the abundance of *A. karroo* and the palatable *Grewia occidentalis*.

Nearby in a similar mosaic thicket region, Andri Marais, Watson and Schmidt (2011) had also observed the feeding of giraffe during one year; they too browsed mainly on *A. karroo* and *A. cyclops*, as well as occasionally on 20 other species. The browsing capacity of the giraffe, judging from the available phytomass of plant species that formed the bulk of their diet at between 2 and 5 m above the ground, was estimated to be between 0.063 and 0.016 BU/ha per giraffe. This was a first approximation of giraffe browsing capacity, but it provides a base for managers to determine approximately how many giraffe an enclosed area can sustain.

Males and females

It is assumed in general that giraffe tend to eat high because they alone of ungulates can reach the leaves of tall trees. In his study of browsing ruminants in Kruger National Park, Johan du Toit (1990b) did find that 90% of browsing by giraffe (*G. c. giraffa*) occurred above the height range of kudu and other ungulates; this enabled them to take advantage of new, protein-rich shoots in the upper canopy. By feeding at heights higher than females could reach, often at full neck stretch, the males also had an advantage over the females, as Masai bulls did in the Serengeti (Pellew, 1983a); Robin Pellew (1984b) found that he could use this difference in feeding posture to identify the sex of a giraffe from a distance. Such a dichotomy suggests the existence, and reduction, of competition between the sexes.

Johan du Toit (1990b) documented this feeding pattern difference between males and females in the Kruger National Park (KNP), South Africa. The mean percentage of time spent feeding above body height (angle above horizontal) was 49% in cows but 84%

Figure 3.6   Bull giraffe are much larger than cows and darker in colour. Photo by Zoe Muller.

in bulls. (This behaviour of eating high is not universal, however; in Tsavo East National Park, Kenya, giraffe in general fed below a height of 2 m about 50% of the time (Leuthold and Leuthold, 1972), and in the Laikipia district of Kenya, most giraffe fed at a height between their shoulder level and their high nose level (Shorrocks, 2009).) Du Toit argues that large bulls spend much time searching for females in oestrus and combatting rivals, so that when it comes to feeding, they do so most efficiently by eating browse that no other animals can reach, rather than searching for nutritious leaves lower down. Feeding with the neck straight up has a negative side, however, because it makes a male less vigilant and more susceptible to lion attack (although Young and Isbell (1991) suggest that feeding high *increases* vigilance). Of 93 adult giraffe killed by lions over a 2-year period in KNP, the cow:bull ratio was 1:1.8, showing that bulls were much more vulnerable than cows (Pienaar, 1969). Cows seldom feed with a full neck stretch and spend almost no time alone.

   Truman Young and Lynne Isbell (1991) delved more deeply into why males and females have different feeding patterns by watching all 53 giraffe feeding on a ranch in south-central Kenya. Once they recognized all the animals individually (although they usually fed in groups), they recorded habitats, the species of each food plant eaten, the height above the ground at which each animal fed, the number of bites taken during each feeding bout, the feeding rate or numbers of bites per second and how long each feeding bout lasted (a feeding bout began when an animal took a first bite and ended when it removed its head from the plant).

The researchers found that groups of females with young were significantly more likely to feed in open habitats with shorter trees such as *Acacia drepanolobium* than were groups of females without young or male groups. The primarily male groups (including a few females) feasted especially on the taller *A. xanthophloea* and *A. seyal* trees. Feeding rates of the adult giraffe were low when they fed near the ground or near their upper limit of feeding and fastest at an intermediate height (about 3.0 m for males and 2.5 m for females). Those in the predominantly male groups fed at the intermediate heights where they fed fastest, but females in the female groups with young fed at significantly lower heights, while males in these groups fed significantly higher than the males in male groups, even though tall trees were scarce in the more open areas these mixed groups frequented.

What to make of these differences? Young and Isbell argue that the sex-based differences were largely the result of constraints on females with young. Females without young hung out in denser habitats than those with young; there they had a greater variety of food plants and could feed at their optimal height. The mothers, by contrast, chose to be in a more open habitat because there they had a better chance of protecting their young from predators; calves are at high risk, with only 30–50% surviving their first year.

Why did males tend to feed high on tall trees? The researchers suggest this is based on male competition for females. Between feeding bouts, the males with females did not lower their heads. By browsing at greater heights, high-ranking bulls can both advertise their presence and status to other males, and increase their vigilance for such possible competition. So both females with young and males likely to mate both adapted their feeding strategies to enhance their reproductive aims.

In the Mikumi National Park in Tanzania, Tim Ginnett (who had earlier worked with the team of Young and Isbell) and Montague Demment (1997) also studied the difference between male and female Masai giraffe in how they ate, what they ate and where they ate. Although adult males weigh more than females (about 1200 kg to 800 kg) and therefore require more food, they spent less time each day actually foraging. This is because they took larger bites (especially during the wet season), they spent less time chewing each mouthful and they spent more time at a food patch, in part because they were tall enough to obtain browse out of reach of females. However, rumination time by day was similar for the two sexes. (In Tsavo National Park, males ruminated for longer periods than did females, perhaps related to the more fibrous content of their diet (Leuthold and Leuthold, 1978b).) It may be that feeding is dependent in part on the number of males and females in a group, but this has yet to be tested. Certainly vigilance behaviour varies with the composition of a group of giraffe because animals stop feeding when they lift their heads to scan their surroundings (see Chapter 4). Management efforts for giraffe should take into account the different foraging behaviour of males and females.

In this same park, Ginnett and Demment (1999) carried out further research not only about direct feeding, but on the preferred habitats of males and females. All the adults fed in the riverine communities where there was ample food, the females increasingly present there during the dry season rather than in the flood plain areas. Females with young tended to avoid the woodland, in the wet season favouring the flood plains and in the dry season, the mixed-shrub habitat. The males rarely visited the flood plains. Females with young presumably preferred open areas because there they did not have to be as vigilant for predators which might attack their calves. This supports the contention of Young and Isbell (1991) that risk is a factor in habitat selection. When females did not have calves with them, they browsed in the same areas as males.

With their further data on feeding, Ginnett and Demment suggest a sexual dimorphism/body size hypothesis relevant to the comparative sizes of males and females. Males are larger (as they are in most ungulates) and are thought to have a greater digestive efficiency, especially on poor-quality forage. Because of their greater size, they will have a lower mass-specific metabolic rate than females and therefore should be able to exist on more abundant, lower-quality forage. Females, of course, require high-quality forage because of their higher mass-specific metabolic rates and the metabolic demands of gestation and lactation.

Other differences between the diet of males and females involve the tannin content of food. Lauren Caister, William Shields and Allen Gosser (2003), anxious to help conserve the West African giraffe in Niger, identified all of the 81 giraffe in the area by photographs of their spotting patterns. Then they observed the feeding habits of these animals from early morning until noon (after which giraffe and people rested in the dry season when temperatures rose often to 49°F). At this time giraffe frequented two areas, the Dallol Bosso (DB) where there are many trees, wetlands, farming and good browsing, and the Intermediate Zone (IM), which is more sandy with scattered trees, low scrub brush and inferior browsing. In these two areas food preferences were calculated by how often giraffe fed on a species and how abundant this species was.

The researchers found that the adults of either sex largely preferred to browse in different areas. The pregnant females tended to frequent DB (64% of females) but no lactating females with newborns were seen there. Females in the IM often had youngsters (42% of females), and sometimes were pregnant (18% of females). Males preferred the DB where *Combretum glutinosum* was a favoured food for both sexes. DB forage is higher in fats, moisture and carbohydrates while lower in ash content than the browse in IM. DB also had the only drinking water, so that females in IM had to trek to DB once or twice a day to drink. Subadults were mostly present in IM feeding on *Prosopis africana*. As well as favoured foods, these various categories of giraffe had an array of plant species that they avoided eating.

Why these choices? Apparently, nursing females feed in the IM because there is less tannin there in the browse that will enter their milk and negatively affect their young. After their offspring are weaned, they return to DB with its superior vegetation. It is impossible to know if this is a peculiarity of the subspecies *peralta*, or only of this population in Niger. For example, earlier data for *peralta* giraffe in Cameroon showed that they preferred *Balanites aegyptica* forage (Ngog Nej, 1984), whereas this species which is common in both areas is little favoured by giraffe in Niger.

Seasonal shifts

Giraffe go where the food is most plentiful and most nutritious, so during a rainy season they will likely migrate to take advantage of new vegetation. During a drought, they will spread out to find what food they can. Vaughan Langman (1978) notes that for the KNP area, when the rainy season ends and the dry season begins (April–May), there is a period before the leaves fall when important minerals such as phosphorous are no longer translocated to the leaves. An imbalance in calcium and phosphorus uptake from forage is detrimental to giraffe, leaving them in poor physical condition by the end of the dry season. Collecting information such as this is important for giraffe populations because unless seasonal migration is taken into account, the animals may be unable to thrive. Here are a few examples.

- In Niger, most of the giraffe population migrate from the same one area to a second each year (Le Pendu and Ciofolo, 1999; Le Pendu *et al.*, 2000). Their time spent browsing on at least 45 plant species was twice as long in the dry as in the wet season (Ciofolo and Le Pendu, 2002).
- In South Africa's Timbavati Private Nature Reserve in Mpumalanga, Anthony Hall-Martin (1974a) noted that during the dry season, giraffe were found in two specific areas while in the rainy season they were much more spread out. There was a significant difference during the two seasons of giraffe presence on *Acacia nigrescens* open tree savannahs and on *Colophospermum mopane* communities.
- In the Willem Pretorius Game Reserve, Free State, South Africa, giraffe move seasonally between the savannah flats and the more vegetated kopjeveld (Kok and Opperman, 1980). In the wet season 90% of their diet comes from only four species: *Acacia karroo*, *Asparagus laricinus*, *Rhus undulata* and *Ziziphus mucronata*. In the dry season they make do with many other less-favoured plants and woody materials. Because of this, as well as osteophagia and the frequent consumption by giraffe of herbs, grass and other low plants, the authors suggest that this reserve is a suboptimal habitat for them.
- In northwestern Namibia, giraffe migrate between three distinct areas, one a year-round refuge, a second for short-term use when vegetation is lush and a third breeding area and sanctuary (Fennessy *et al.*, 2003).
- Female giraffe in the Etosha National Park, Namibia, changed their behaviour significantly in the dry season compared to the wet season: with little or no rain, herds were smaller in size, the number of female associates was down and the strength of these associations was reduced (Carter *et al.*, 2012b). In the dry season as well, the females had poorer body condition and foraged farther from each other.

4 Social behaviour and populations

For reproductive behaviour and behaviour in zoos, see Chapters 9 and 10.

How social animals behave among themselves is of great interest to behaviourists, but also to those who oversee wildlife populations (Carter, 2009). For example, if a contagious disease is present in one individual, how that animal interacts with others will help a manager figure out how the disease might spread, and therefore how it might best be combatted. Or if a few individuals are going to be transported to another area, picking giraffe who are friends to travel together will ease this process.

Now that we know from DNA samples that most subspecies of giraffe have been isolated from each other for very long periods of time, and therefore may actually be separate species, it would be exciting to find whether these 'species' have unique social behavioural patterns too. This may be true, but we won't know this any time soon. There are far too few studies of behaviour, each of which is incredibly labour-intensive and expensive. As well, it has already been shown, as we shall see in this chapter, that behaviour is shaped in part by local environments and by group composition. In this chapter it is assumed that giraffe of all races behave in more or less the same way.

Giraffe herds

The name 'herd' often applies to a group of animals which stay together for many months. Giraffe herds are different, in that individuals come and go day by day as the spirit moves them; sometimes a herd of 100 was seen many years ago, and often one or only two together. Recently we have learned that giraffe live in fission/fusion societies similar to those of chimpanzees and human beings, but the first person to study this question, Bristol Foster, a former university classmate of mine, could not know this.

Foster made a detailed study of giraffe herd sizes and compositions in the Nairobi National Park (NNP) in the 1960s. He pioneered the system of using photographs of each animal so that he could recognize Masai giraffe individually who visited the Park during a study of over 20 months. Eventually he had records for 241 animals, any of whom he might see as he drove every week through the Park, noting down all the individuals he encountered.

Foster's method of identification of individual giraffe by photographs or drawings of coat patterns was followed by researchers for nearly 50 years. Then Bryan Shorrocks and Darren Croft (2009) devised a scheme whereby the neck patterns of reticulated giraffe can be coded into an Excel spreadsheet which allows several hundred individual giraffe to be recognized and located within a database. Their software package using Ucinet for Windows further enables social network patterns among individual giraffe to be documented, which will be a valuable research tool in the future. Similar important work is being done by Lindsey Stutzman and Elizabeth Flesch (2010) for Thornicroft's giraffe in Zambia.

More recently, Douglas Bolger and colleagues (2012a,b) have developed a software program that can easily distinguish many hundreds of individual giraffe in the field from the pattern of their coats. From over 1200 images taken in the field over a long period, they were able to identify over 600 unique individuals of Masai giraffe. Using this technology, Derek Lee and Monica Bond (2011) have launched a multi-year intensive study of Masai giraffe populations and Giraffe Skin Disease epidemiology in the Tarangire–Manyara ecosystem. Another system, ExtractCompare, uses a three-dimensional model which can match images of the same individual taken from different angles (John Doherty, 2011). With it, researchers can make use of suboptimal photographs from a variety of sources.

Friendships within a herd

Foster's (1966) definition of a giraffe herd as a collection of individuals less than a kilometre apart and moving in the same general direction is still useful, but gradually over the years researchers have been able to define herds more precisely. Long ago, Bristol asked me to help him analyse the mass of data he had collected from the NNP which I was happy to do (Foster and Dagg, 1972). The results startled us. He had records of many groups seen together in his weekly forays through the Park, but the evidence showed that they were rarely the same individuals. Social animals in general go about in groups almost by definition, but the groups usually comprise the same animals. For giraffe, it seemed that a herd wasn't really a herd at all, but just a collection of individuals who happened to be in the same place at the same time, perhaps taking advantage of an enticing patch of food. The next week most of them would be gone, replaced by other wanderers.

The data showed that:

- adult males tended to be loners, rarely seen with other males, females or young;
- females were usually in groups, but the group composition changed continually. Even when two individuals were seen together a few times, these associations were usually not successive. On average, an adult female was seen with another particular adult female 3.4 times during the long-term study; and
- with the exception of mothers and their own calf, particular females and young giraffe or two young animals were seen together only 2.8 times during the study.

Barbara Leuthold (1979) who researched Masai giraffe in the Tsavo East National Park found that their social behaviour was similar to that of the NNP group:

- groups were usually small (mean average 3.8 animals);
- calves might or might not belong to nursery groups;
- subadult males tended to hang out together, travel in a group and spar often;

Figure 4.1 Bristol Foster measures the height of 5.5 m for a male Masai giraffe foraging.
Photo by Bristol Foster.

Figure 4.2   Large congregation of Rothschild's giraffe. Photo by Zoe Muller.

- adult males were loners who walked long distances to find females in oestrus with whom to mate and drove away other males from these females. Leuthold felt that the adult males all knew each other and were therefore aware of who was the most dominant because in the past he had won most battles against the other males;
- old dark-coloured males were loners who neither wanted to mate with females nor to travel much, all passion spent; and
- females were usually with their young and/or together with other females and young, but they frequently changed their daily pals. However, unlike in NNP, the young were more often with their mothers: 15 known young associated with their mothers for periods from 12 to 16 months, and in one case for 18 months.

Twenty years later, Yvonnick Le Pendu, Isabelle Ciofolo and Allen Gosser (2000) published a study of the social organization of *G. c. peralta* giraffe in Niger. They too might have found different social behaviours than the above because of their animals' desert-like environment, but the giraffe there also did not have strong preferential associations even though there were at most 63 of them – individuals (other than mothers and their young) associated with only half of their partners of the day before. Groups were smaller in the dry season (six animals on average) when food was harder to come by than in the rainy season (nine animals) when it was abundant. Of the solitary individuals, 75% were adult males (most of them over 10 years old), 21% adult females and 4% juveniles. As in other populations, the subadult males were the most interactive, mounting, sparring and necking with each other. Individuals mingled freely with those of other classes, ages and sex.

In 2009, Fred Bercovitch and Philip Berry (2009a) solved the problem of friendships within a herd when they published a long-term survey covering 34 years of data for Thornicroft's giraffe in Zambia. Their findings are exciting because they predict a fission/fusion society rare (and until fairly recently unknown) in wild species. This type of society involves the formation and dissolution of subgroups within a larger social network based upon preferential associations. Now we know that infrasound (sounds too low for human ears) occurs among giraffe and other herd animals, even if the members are physically far apart and out of sight of each other (von Muggenthaler *et al.*, 1999). Maybe all the giraffe in a large area form, in effect, a large herd, with members simply moving about at will within their large community? Perhaps they even keep track of each other, as people now do with their cell phones?

Between 1963 and 1969, Philip Berry (1973) had documented that an average giraffe herd size in the Luangwa Valley of Zambia numbered four animals, with 25% of herds containing six or more individuals. Many years later, Bercovitch and Berry (2009a,b) reported on Berry's findings over a long time period and on much more. In 2007 the average herd size was 3.6 animals, with 21% of herds having 6 or more giraffe. Not much had changed. Herd size depended on a number of factors: smaller in dry seasons and in woodland areas as expected, but also on four other aspects.

(a) *Foraging strategies.* Giraffe eat over 90 species of plants in the South Luangwa Valley and dozens of species in other parts of Africa. There is no competition for food, as any one animal can munch away at any plants he or she fancies. They do not drink every day, so seldom gather at a river or pond. If several animals eat in the same area at the same time, this is a matter of chance. (In contrast, in Namibia, access to food and water is limited so that numbers may gather at a rich food source or a river and appear as a large herd (Fennessy, 2004).)

(b) *Reproductive strategies.* Various studies indicate that bull giraffe frequent dense forest and thicket areas more often than do females, and that males are more often alone, presumably travelling to find a female in oestrus with whom to mate. These different activities will also impact herd size.

(c) *Social strategies.* Long-term studies show that related animals such as mothers and their daughters may stay more or less together over the years. With animals travelling about continually to different feeding areas, they routinely run into their neighbours; it seems certain that they are familiar with all the animals around them, even if they are usually apart. Apparently the small number of giraffe usually seen together are really subunits of a much larger social group perhaps numbering up to 30 or 40 animals; giraffe seemed to be a fission/fusion social species like chimpanzees.

(d) *Antipredator strategies.* Herds in open areas of Zambia were about 1.5 times the size of those in thickets, where giraffe are more vulnerable to predation by lions. Gathering together in open areas where lions can be seen is a good idea because there are many eyes to detect them.

All of these factors affect the size of herds of giraffe. Together they strengthen the idea that giraffe do indeed live in a fission/fusion social society. Other data for the Thornicroft's giraffe (common to other subspecies as well) (Bercovitch and Berry, 2009a,b) are:

- large bulls usually travel alone;
- cows are seldom alone;
- young are born any time of the year;
- herds tend to average from 3 to 5 animals;
- 20–25% of herds consist of 6 or more animals; and
- herds nowadays seldom have over 30 giraffe.

In a second paper elaborating on their findings about herd size, Bercovitch and Berry (2013) set the giraffe firmly as one of at least nine mammalian species known to have fission/fusion societies, many of which have been classified as such only in this century. They include the African elephant, spotted hyena, dusky dolphin, sperm whale, killer whale, hamadryas baboon, chimpanzee, spider monkey and now giraffe. The authors believe that the fission–fusion social systems evolved in response to spread out, unpredictable, restricted and ephemeral food sources; when some members of a group came upon the food, they would broadcast this finding to kin and others in their group so they could join in the feast.

These fission–fusion species have many of the same attributes as do giraffe:

- possible communication about sources of food (in giraffe via infrasound);
- non-seasonal breeding;
- large home ranges with spread-out food sources;
- flexible and labile diets;
- cooperation and/or tolerance in offspring rearing (for example, a giraffe mother may feed 1 km away from her calf if it is near other giraffe and seems safe);
- sexual segregation; and
- subunits usually containing both kin and non-kin (if this can be verified) with fewer than 10 individuals.

Fission/fusion groupings also help explain the social networks of 535 known Angolan giraffe living in the Etosha National Park, Namibia. Kerryn Carter and colleagues (2013) analysed the composition of giraffe in the 726 groups that were sighted, photographed and recorded over a 14-month period from vehicles driving on the Park roads. The genetic relationships of females were obtained by collecting 70 skin biopsies for DNA analysis. (The animals darted by a remote biopsy dart delivery system seemed unaffected by the process, usually resuming foraging within a few minutes of the dart being discharged.) The researchers also took GPS readings so they would know which giraffe used specific areas.

The males didn't associate to any extent with other males (and male/female associations were ignored because they were chiefly driven by sex), but many females were observed either moving about with, or not associating with, other individual females. Pairs often browsed in the same areas, and those exhibiting preferred relationships (because they were especially often seen together) were more closely related than would be expected by chance. However, only about one-quarter of the variation in female pair close associations could be explained by spatial overlap and relatedness. Many twosomes apparently hung out together because they liked each other.

The concept at least of fusion is noted in an unusual finding in Arusha National Park, Tanzania (Pratt and Anderson, 1982). Strangely, even though this park is small, the giraffe made it seem even smaller by having two separate populations, North and South. The researchers made 6836 sightings of cows and young animals, but a North individual was virtually never seen in the South, or a South one in the North. Fourteen bulls did cross this psychological divide, but these were only 3% of the total population. Both populations were aging, with a low birth rate and a high mortality of young during their first year of life, even though there were leopards and hyenas but no resident lions in the Park.

Long-term bonds

Are there long-term bonds between two giraffe? Early researchers could not be sure. For a mother and her calf, these bonds are either thought to disappear early (du Toit, 2009b), to become broken after 14–16 months when a new sibling is born (Langman, 1977) or to persist for 22 months or for years (Pratt and Anderson, 1985). Some research indicates that there are strong bonds between cows (Pratt and Anderson, 1985; Fennessy, 2004), between individuals of about the same age (Fennessy, 2004) or between mother and mature daughter (Berry, 1978; Saito and Gen'ichildani, 2011 – for females in Katavi National Park, Tanzania). For wild females in Etosha National Park, Namibia, Kerryn Carter and colleagues (2012a) found that some associations between pairs of giraffe were strong, lasting for up to six years. As young females reach adulthood, they associated with increasing numbers of females, thus increasing their social network.

From their thorough analysis of which two giraffe were most closely involved with each other, Bercovitch and Berry (2013) determined that cows and their offspring, as well as specific pairs of cows, formed close associations. Only 11% of female–female pairs did not form a herd with each other, while 30% of male–male dyads were never observed together in a herd. While 57% of kin dyads had a strong social association, only 7% of non-kin did. Every possible sister–sister dyad was observed within one herd, while four brother–sister dyads were never recorded within the same herd. Adult males were solitary 70% of the time, and only 4 males were ever seen in a herd with their mother over the 34-year study. The authors dismiss categorically the idea that herds are associations of individuals with no close ties.

Bulls

David Pratt and Virginia Anderson (1982, 1985), who researched giraffe social behaviour in three different Tanzanian national parks over 3 years, described typical behaviour of bulls and cows. They paid special attention to mature bulls in Arusha National Park which they classified as A (dominant), B (subdominant) and C (younger least dominant

Figure 4.3   Rothschild's mother nuzzles her young calf. Photo by Zoe Muller.

adult males). The C bulls who were at least 5 years old were rarely alone, but usually in close groups of other males, females and young giraffe. In Tarangire these young bulls spent two-thirds of their time in mixed groups of giraffe (with females, young and/or other bulls) and more than one-quarter of the time in all-male groups. As they grew older, their necks became heavier and their horns longer and thicker. Sparring was a common activity of young bulls, and often among those of different sizes, instigated by the smaller animal. John Doherty (pers. comm., 2012) who studies reticulated giraffe in Samburu, Kenya, notes that the physical closeness of young subadult males who hang about together and often spar are rather like young men with time on their hands, fooling around with nothing better to do.

The sparring of young males seems more like a stately dance than a scuffle, with two animals standing side by side, head to head or head to tail, taking turns to gently hit each other's bodies with their horns (Innis, 1958). Often they pause at length between hits or others join them in a group spar. Sometimes they conclude their sparring by browsing side by side, perhaps one with his nose straight up in the air. Sometimes one mounts the other. Such activity seems to entice other males to spar too (see Chapter 5).

Unlike these sociable younger males, the larger and stronger A and B bulls were often with females or young, but they shunned their peers. Pratt and Anderson (1982)

Figure 4.4   Subadult males (here Rothschild's giraffe) usually hang out together. Photo by Zoe Muller.

never saw groups of only A bulls, or more than two A bulls together. B bulls were more likely to be together, but were much less social-minded than C animals. The A bulls who were known to be dominant had presumably gained this status when younger because they were never seen sparring with other males, as B males did sometimes and C males often did. When an A bull joined a group, he would approach the B bulls and they would depart – a look was enough to unnerve them. Each population had a dominance hierarchy among the mature bulls so there was no need for them to fight as they had when they were younger. Pratt and Anderson (1985) noted 129 such displacements of one usually smaller bull by a larger one. Usually the larger of two A or B bulls was the more dominant, but not always. They only saw two serious fights, one in which a B bull clubbed a C bull several times before the latter ran away and in the other, when a B bull who had been courting a young cow for some hours viciously attacked an approaching bigger B bull with violent blows in rapid succession which drove him away. In alphabetical order the bulls were especially attentive to cows, flehmening them, following them if they were in oestrus for days if necessary and mating with them. If a male joined a group with calves, he typically bent down to touch in turn each small animal's head with his nose.

In the Etosha National Park, Rachel Brand (2011b) found the above mating pattern especially obvious because the less dominant males were noticeably paler than the dark older males, and it was the latter who were the most likely to attract the attention of females and to mate with them. It seemed obvious that mating success was highly skewed in favour of a small number of dominant, very dark males.

Cow herds

Large bulls are most often seen by themselves, while cows are seldom alone. They team up with other cows who happen along, and of course with the young born to any of these females. In their 10-month study at the Tarangire National Park, Pratt and Anderson (1985) recorded 112 instances of one cow nosing or rubbing against another with her head. A cow bumped another or hit her with her head 12 times, pressed her chest against another's rump 7 times, mounted one time and urine-tested once. The cows in the Serengeti and Arusha National Parks had had similar amounts and types of cow–cow contact (Pratt and Anderson, 1979, 1982), but there were many fewer such interactions between cows in Tsavo East National Park (Leuthold, 1979).

In Tarangire, the most stable groups were those of one or a few mothers and their young who were quite often seen together, the calves especially forming bonds with their play. Pratt and Anderson (1985) were able to quantify the relationships of one cow, Canopener, with her seven closest buddies. They saw her 109 times, and usually more than half of her group was with her in a variety of combinations. Often the group did not seem to have a leader, but Canopener took this role 49 times, with others usually following her. She was, in fact, the central figure of the group. The leader of a herd of giraffe, if there is one, was almost always a middle-aged or old cow, usually with a calf; this is true in other ungulates too, such as red deer/elk, sheep, goats and mountain goats (Dagg, 2009).

Day care?

Vaughan Langman (1977) carried out a study of relationships between mothers and their calves in South Africa, bringing entirely new information into the equation. My study group in South Africa had had few calves (Innis, 1958) and Bristol Foster's (1966) Kenyan programme hadn't focussed on the young (although he did note that babies and their mothers were often apart during the days when he surveyed his giraffe, whereas older calves and females tended to be together).

Langman noted that newborn giraffe stayed hidden in thickets or tall grass for their first 3 weeks where their mothers came three or four times a day to nurse them for a few minutes. Then the pair joined one or more other mothers and their young in a nursery group, with the young remaining in an open area often on high ground with tall grass during the day. Sometimes a female stayed with the calves while the other mothers went off to feed, making in effect a crèche. Because they are often in the open, young calves may be spotted and frightened by a predator or a human observer. In that case they run off in a group, but circle around to come back to their starting point where their mother left them. If there is a mother giraffe with them, she will go to her calf if she is uneasy and nudge it to make it follow her. The other calves, sensing her nervousness, will follow after both of them. At night, one or two females stand watch near the nursery group, alert for the approach of predators; they change shifts during the darkness so that no one female is

Figure 4.5 Rothschild's calves, born the same day, are 2 weeks old. Note red-billed oxpeckers and umbilical cord remnant. Photo by Zoe Muller.

on duty all night. The young were weaned at 6–8 months but remained with their mothers for at least 14 months. The young began to nibble at leaves when they were quite young but did not ruminate until 4–6 months old.

Langman (1982) gained further information by putting radio collars on a series of animals of different ages. They were specially designed: the transmitter had to be strong enough to stay on the collar even if the animals pushed through vegetation or tried to dislodge it; it must not rub on the skin of the giraffe and hurt it; and each should be made to fall off after 2 years, which Langman hoped would happen by using bolts that would wear out at that time.

Observations of Pratt and Anderson (1979) on mothers and their young in the Serengeti were similar to those Langman had noticed in South Africa. If there was food available, the mothers fed nearby when their calf was very young, presumably keeping a watch for predators such as lions, hyenas or cheetahs. Each was attentive to her newborn as she nursed, touching, rubbing, sniffing and licking it, keeping other giraffe away. The baby in turn licked, rubbed and touched its mother. The calves groomed and became playful at 2 weeks, gambolling about if their mother was near, but most of the time they lay quietly. In this way they did not draw attention to themselves and conserved energy that would enable them to grow quickly until they were too large to be prey for most predators.

By contrast, when Pratt and Anderson (1982) turned their attention to the 471 giraffe in the Arusha National Park in Tanzania, they found no crèche systems, although there were at least 22 calves in the relatively small, heavily vegetated area. However, the Park is surrounded by agricultural lands so that the adults could not venture far to new feeding areas. There, neonates (calves less than a month old) stayed close to their mothers – in only 6 of 90 sightings were the two apart. The one calf Pratt and Anderson saw lying away from its mother soon disappeared, presumed dead. Calves were rarely left unattended in this Park where water is plentiful, unlike Serengeti calves whose mothers had to travel far to find it, especially in dry seasons (Pratt and Anderson, 1979). Both mothers and young initiated nursing but, as in the Serengeti animals, the mothers terminated almost all of these bouts, often when other giraffe tried to suck too. Even though the nursing sessions were not lengthy (the 49 sucklings seen in the Arusha group lasted on average 59 s compared to 66 s for the Serengeti giraffe), unsuccessful attempts to suck outnumbered successful attempts. The oldest calf to suck was 9 months in Arusha Park, but 22 months in the Serengeti.

The calves which were usually with their mothers and not in crèches as in other areas did, however, form what might be called play groups, with up to five seen together fairly often (Pratt and Anderson, 1982). This was because their mothers tended to hang out in the same area, where most of these calves spent more time with their peers than with their mothers. They stayed nearer to each other and had lots of physical contact: nosing, rubbing, sniffing, licking, kicking and gambolling. Males and females played together, but such activities waned sharply with age, from 0.65 bouts per hour in the first week of life, to 0.17 bouts in the second month and 0.08 bouts after one year. Young calves also like to hang out together in zoos, mirroring the crèche system sometimes present in herds of wild giraffe (Dagg, 1970b; Rose, 2011).

Taping of behaviour

A new tool in the arsenal of behavourists is that of setting up cameras in the wild so that the presence of giraffe and their activities can be recorded without people being present to perhaps bias the goings-on. Such taping provides information about the social and individual behaviours of giraffe, their travels and their distribution. In the Etosha National Park in Namibia, cameras set up at 4 and then 8 sites since 2010 have captured over 7000 pictures of giraffe doing their thing (Dean, 2012); this is a splendid place for taping, given that the vegetation is not high enough to impede views. The images documented the composition of giraffe herds, mating, suckling, drinking water, feeding and chewing of bones.

Population densities

Forty years ago, accurate data for population densities of giraffe in the wild were lacking. There were reasons for this.

(a) Censuses of giraffe in an area were usually taken by aeroplane. Their accuracy was necessarily approximate because of variables such as height of the surveying plane, flying time, number of observers in the plane, weather, time of day and dense vegetation.

(b) It was impossible to delimit the area these giraffe inhabited. Giraffe follow the food, so their spatial distribution is constantly changing along with seasonal changes of vegetation. For example in the hot-dry season near the Ombonde River in north-western Namibia when favoured *Faidherbia albida* pods were available, giraffe density was 3.62 per km^2. However, there were no giraffe there in the wet season and 0.59 giraffe per km^2 in the cold-dry season (Fennessy *et al.*, 2003).

(c) Other factors also affect their distribution – sources of water, other herbivorous species, predators and elements that might prevent giraffe from leaving or favour giraffe immigration.

Densities of giraffe are easy to tabulate, given that one need only, for example, divide the number of animals believed to be in a park by the area of the park, but they are unlikely to be completely accurate.

Densities calculated on the ground rather than from the air are more exact, as the researchers usually know the terrain and the habits of their animals. For example, Bristol Foster (1966) collected long-term data for Nairobi National Park where the density of giraffe remained constant at 0.72 per km^2 (2 per mile^2) for at least 7 years, even though during that time individuals continually wandered into the Park and out again as the spirit moved them. If the average weight of giraffe was 800 kg each, then the biomass was 576 kg per km^2, a value that would vary should lion predation or climate change substantially.

With the proviso above for most density calculations, and because animal densities are useful even if not exact, Dagg and Foster (1982) present a chart of 30 densities in a range of habitats from 5.3 giraffe per km^2 in part of Timbavati, Mpumalanga, South Africa, to 0.01 giraffe per km^2 in the desert region of Baragoi, Kenya. (The greatest density was of 14.1 per km^2 in Wankie National Park (then Rhodesia) main camp area, but this was far from a normal situation.) The density of giraffe at Timbavati was well above the carrying capacity of the area because later, during the dry season of 1965 following a period of low rainfall, 182 giraffe (over 15% of the population) died from malnutrition and starvation (Hirst, 1969). At this time S. M. Hirst reported increased predation by lions.

The density of giraffe in Africa is least where the vegetation is most sparse – in desert areas. The lowest recorded rate is in northwestern Namibia where there are 0.01 giraffe per km^2 (Fennessy, 2009). In northern Botswana, the density of giraffe is 0.08 giraffe per km^2 (Chase, 2011).

Depending on conditions, density values can change dramatically. In 1995, the density of giraffe in Niger was also 0.08 per km^2 (Leroy *et al.*, 2009); this was not because the vegetation was limited, but because the giraffe were being slaughtered. As described in Chapter 11, the giraffe in this country were at high risk in the mid 1990s, with only about 50 animals remaining. At that time every effort was made to protect them – poaching was prevented and the giraffe were free to breed in regions with substantial suitable vegeta-tion where there were no predators. With the killing stopped, their numbers increased dramatically thanks to careful management, making a world record for population growth (Suraud and Gay, 2010). The rate of annual increase was an astonishing 19% (1996–1999) and 13% (2005–2008) (Suraud *et al.*, 2012). Efforts have begun to use a suite of microsatellite loci to potentially allow a genealogical or pedigree tree of these animals to be constructed.

An example of a decreasing population comes from Arusha National Park. At the end of their year's study from 1979 to 1980, Pratt and Anderson (1982) recognized by their spotting patterns 462 individuals, 242 females and 220 males, almost the entire population. During the year 22 calves were born, but only 13 survived, even though there were no resident lions in the Park. Yet the expected number of births of young for the 172 cows (judging from other populations) was 103 calves per year. This population was obviously senescent, with calves and juveniles comprising only 24% of the population, so the number of giraffe was on the decline. Presumably this was because the Park was surrounded by agricultural areas so that giraffe could not emigrate to new regions as forage became scarce. (The researchers note that during nights along the road going north and south out of the Park, often a small group of giraffe of all ages travelled up to 7 km, returning to the Park at daybreak. Were they trying to scope out new territories with a view to possible emigration?)

Distribution and numbers of giraffe in the wild have been documented from aerial counts, vehicle counts, 'mark–recapture' sampling and camera traps. It would be valuable for researchers to compare these various surveying methods in the field to determine which are the most accurate and cost-effective.

Sex ratios

Birth records from zoos indicate that, with a few notable exceptions, about as many males are born as females (see Chapter 9), and Arusha National Park had equal numbers of adults too – 176 males and 172 females (Pratt and Anderson, 1982).

However, because adult males and females usually go their own ways, the sex ratios can change depending on the area. For example, in northwestern Namibia for the three study areas set up in the Hoanib River catchment, each had a different sex ratio (Fennessy et al., 2003):

Lower Hoanib River sex ratio	1 male to 1.38 females
Hobatere game park	1 male to 1.6 females
Ombonde River	1 males to 0.62 females

The sex ratio of giraffe was also significantly skewed for Masai giraffe in the Serengeti National Park, Tanzania, but for a reason that had nothing to do with vegetation. Wilfred Marealle and his team of four researchers (2010) monitored the giraffe there for 2 weeks every month over the period of a year, 2008–2009. They took photos of the individuals they encountered and noted the sex ratios of 2947 giraffe. Their data were pooled into three age groups: females were 62% of adults, 54% of subadults and 67% of calves. These calves were born in every month, with a low of 8 in February and a high of 27 in March.

The team also made an assessment of the behaviour of giraffe when an inquisitive research vehicle first encountered a group. Were they (a) feeding, (b) vigilant – either standing and looking around or escaping, or (c) acting in other ways such as walking, mating, resting or nursing? Then they recorded the flight distance of each group. When giraffe were first sighted, the car was stopped. After all the needed information had been recorded, a single person left the vehicle and began walking steadily toward the animals. The distance of the giraffe from this person when they turned to flee was then measured.

For research purposes, the Serengeti had been divided into two areas: (a) at high risk of illegal hunting where villages adjacent to the Park were known to poach bush meat, and (b) at low risk farther from such habitations. In areas of high risk there were 67% more females than males, while in the areas of low risk there were 55% more females. Giraffe were more vigilant with a longer flight distance in high-risk areas and less vigilant (more likely to be feeding) in low-risk areas.

These findings indicate that illegal hunting was taking place in high-risk areas where the larger adult males were targeted more than adult females. Subadults which have less meat on them than adults were presumably less likely to be killed, as the sex ratio was much less skewed at 54%. In addition, lions tend to kill males more than females, presumably because males are more likely to be alone than are females and to inhabit forested areas where lions cannot be seen in the distance (Owen-Smith, 2008).

The sex ratio of the calves presented a different picture altogether. The proportion of female calves was very high, especially in areas with high risk of illegal hunting. The plausible explanation is that the adult females there were giving birth to more females than males, a possibility present in many species of mammals, perhaps connected with the role of glucose in reproductive functioning (Cameron, 2004). (Even in human beings, mothers tend to give birth to relatively more daughters than sons after catastrophes such as the terrorist attack of September 11, 2001 (Catalano *et al.*, 2005), phenomena correlated with high levels of stress hormones (corticosteroids) in the women.) Presumably giraffe in high-risk areas must be constantly aware, and stressed, over disturbances by illegal hunters checking snares, stalking game and removing dead bodies from the area of a variety of species including giraffe.

Age ratios

The population of giraffe in Nairobi National Park monitored by Bristol Foster (1966) had a consistent density over a 6-year period. His collected data, therefore, give a rough idea of what the age composition of an average population, neither growing nor shrinking, might be:

    31% adult females,
    25% adult males,
    12% young in their fourth and fifth years,
    8% young in their third year,

10% young in their second year,

14% in their first year (of which 5% were less than 3 months old).

These figures indicate that mortality was greatest in the youngest animals, that death probably occurred in every year thereafter for subadults and that adult females lived longer on average than adult males.

We know that male and female giraffe often choose to forage in different habitats (Chapter 3), and individuals of different ages may do the same, as Julian Fennessy and his colleagues (2003) document for giraffe in northwestern Namibia. For three different types of vegetation, their findings were:

	Adult : subadult : juvenile		
Lower Hoanib River: year-round refuge	10.7	: 5	: 1
Ombonde River: short-term use	8.8	: 2.8	: 1
Hobatere game park: breeding area/sanctuary	2.8	: 1.2	: 1

Transporting giraffe

In Africa some regions are judged to be in need of giraffe, while others are thought to have too many. In either case, managers may decide that some should be transported to another location. This is described in Chapter 11 for several small groups of endangered *rothschildi* giraffe to increase the chances of their race surviving. A larger translocation was carried out for reticulated giraffe from the Aberdare Country Club sanctuary in Kenya where the population was considered to be exerting unhealthy pressure on the habitat (Chege, 2012); there were 43 giraffe there, although the recommended number was only about 15–20. The transported individuals were taken to the Sera Wildlife Conservancy which wanted to increase its number of wildlife species.

The operation was carried out in August 2008 by a team of 24 people. A makeshift plastic fence was set up in a funnel shape leading to the water and a salt lick so that giraffe could be surrounded and become used to this smaller area for a week before the actual capture occurred. Here they were fed lucerne and dairy cubes to supplement the limited browse available. On each of the 5 days of capture, giraffe were herded forward onto a truck by people, vehicles and, once, a helicopter; the truck then drove the 240 km to a temporary holding facility in their new home. A total of 26 animals were captured, although 3 died en route – 2 fell during transit and one was electrocuted at a river crossing. The mortality rate of nearly 11.5% was thought to be acceptable because it would likely have been much higher if the animals had had to be anaesthetized for the journey.

When moved to a new area, the transient giraffe should be kept in quarantine for a period, usually to protect resident animals. They should be monitored for disease through the taking of blood samples, given vaccinations for rabies and tetanus if appropriate, and have their faeces examined for parasites.

Predators

Like most prey species, giraffe are suspicious if they encounter something new. Vaughan Langman (1977) reports that a giraffe stops about 100 m from a strange, possibly dangerous object, stares at it intently, then gives an audible hiss through its nose. After a few minutes it moves a short distance sideways, in an arc, to repeat this performance. It may do this 15 times before being convinced that the object poses no problems and can safely be ignored.

Lions are the sole predators of large giraffe, but calves may be taken by cheetahs and hyenas as well (Pratt and Anderson, 1979). U. de V. Pienaar (1969), who carried out an extensive study of predation in Kruger National Park from 1933 to 1946 and 1954 to 1966, reported that during these periods 675 giraffe of all ages were killed by lions, 2 by crocodiles, and 2 calves were taken by cheetahs. In 1967 and 1968, 108 more giraffe were killed by lions (64 adults, 18 subadults and juveniles, 12 infants and 14 of unknown age), so it seems that lions kill giraffe at random as encountered, rather than going after members of one particular age group.

Giraffe weakened by poor forage or not enough of it will be at special risk of becoming food for lions. However, males are more likely to be killed than females as they are often on their own rather than in a group, and in dense vegetation where they are at a disadvantage because they cannot easily spot and flee their attackers. But again, solitary bull giraffe males sometimes hang out in open areas with other ungulate groups such as zebra, eland, wildebeest and ostrich (Ziccardi, 1960); presumably their tall neighbour is especially likely to spot predators and he, too, would benefit from their many eyes and any stampedes.

In the Kruger National Park, the only other ungulates more favoured and obtainable by lions as food were waterbuck (preference rating 6.05), kudu (3.82) and wildebeest (3.06). The preference rating for giraffe of any age by cheetah was 0.14, the lowest for any of their prey species. Pienaar calculated the preference rating of lion for giraffe as:

Preference rating = kill percentage divided by percentage abundance = 3.01

Large numbers of giraffe calves would very likely also have been killed by predators, but were not included in these tabulations because their carcasses are small and quickly devoured.

Giraffe meat may be tasty for lions, but they do not kill all that many (small calves excepted). In one report of predation in Serengeti National Park, over a period of 8 years there were 125 lion kills, but only 6 were of giraffe; these had all occurred in bush areas rather than in the open (Kruuk and Turner, 1967). In the Murchison Falls National Park in 2003 and earlier, lion kills of giraffe had not been encountered at all (Brenneman *et al.*, 2009a).

Giraffe fight back against predators with a strong chop or strike from a front leg or a powerful kick from a hind leg. YouTube has a video of the mother of a newly killed calf kicking out at the lions responsible. Some giraffe have killed lions with their blows. B. Putnam (1947) recounts a lion stalking a group of six giraffe that included a calf. The other giraffe fled, but the calf was too slow to follow so its mother tucked it under her

Figure 4.6   Megan Strauss on an aerial survey of Masai giraffe in the Serengeti. Photo by Daniel Rosengren.

body, then wheeled to face the lion as it circled around them at a distance of 30 m. If it came closer, the mother struck out at it, forcing it back. After an hour, the lion gave up and left. Other tales of giraffe and predatory dramas are cited in Dagg and Foster (1982, pp. 38–39).

Megan Strauss and Craig Packer (2013) report on a 'lion claw mark index' which will be vital in documenting scary ordeals of giraffe which no people were present to document. On the bodies of giraffe she had photographed, Strauss found she could detect lion claw marks which were easily distinguishable from other scars or skin ailments. The bodies of 702 individually known giraffe in the Serengeti National Park revealed the following:

(a) for giraffe older than one year, 13% had lion claw marks on their hide, a number surprisingly high when attacks themselves are rarely observed;
(b) claw marks were most often seen on the sides or rumps of giraffe, but occasionally on the neck or a leg too;
(c) no marks were seen on young giraffe, suggesting that those attacked rarely or ever survive;
(d) the prevalence of such marks varied widely in three different areas; and
(e) even the largest males are at risk of lion attack.

In the future, Strauss plans to analyse:

(a) how the claw mark index correlates with actual mortality rates for adult giraffe;
(b) the relationship between claw mark presence and lion density; and
(c) how this index relates to other prey species and other ecological and environmental variables.

By far the most dangerous predators of giraffe are human beings who, largely because of poverty, hunger and greed, kill many thousands of giraffe every year. Ignorance is also involved because, for example, sometimes rumours are spread, probably by poachers, that someone with HIV/AIDS can be treated by eating the bone marrow or brains of a giraffe (Strauss, 2009).

This huge problem of poaching has been well documented in Tanzania where the giraffe, the national symbol, is one of the few mammals not available for recreational hunting. In the Serengeti ecosystem giraffe declined from an estimated 10,600 in 2003 to an estimated 5200 in 2006 (Strauss, 2009; Strauss and Kilewo, 2011); across the border in Kenya's Masai Mara National Reserve, there was a similar decline. The *Arusha Times* reported in 2008 that 210 giraffe had been slaughtered for meat, hide and tail hair in a 10-month period in a corridor west of Kilimanjaro, while in the Katavi–Rukwa ecosystem of western Tanzania poachers remove as much as 40% of the giraffe population each year (Strauss, 2009).

Snares set by poachers cause immense pain and usually death to animals that become trapped, with no benefit to the poacher if the animal is strong enough to break loose with the snare embedded in its neck or leg. Even snares near the ground are dangerous, as giraffe often browse on low bushes (Anon., 2012c). Once a giraffe is caught, it can suffer for months as the wire works into and infects its flesh. One young female giraffe living in the Mbuluzi Game Reserve in Swaziland limped badly because a cable snare had become embedded deep into the flesh of her fetlock joint. Rangers tried to save her, but the wound was so serious and painful that they felt impelled finally to euthanize her. The female would have produced much needed calves for the reserve had she lived, so the rangers were especially devastated by her death.

In the Serengeti, snares for giraffe are often of heavy-gauge cables set high in trees among the vegetation (Hoare and Brown, 2010). If a giraffe is caught by the neck, it may circle the tree frantically trying to free itself, in the process becoming entangled ever tighter until it dies an agonizing death of asphyxiation. If it breaks free, it may suffer for months from lacerations or fatal blood poisoning or, if it is lucky, it may be spotted by rangers who arrange for a veterinarian to free it. This involves darting the animal with an anaesthetic, following it to bring it down as it staggers about (other species are felled as soon as the drug takes hold) and then holding it down as it is quickly given a reversal drug, restrained and the snare cut off. Before it is allowed to get up, its wounds are treated and it is injected with an antibiotic. Megan Strauss has recently inaugurated an index of snaring events based on marks left by snares on the skin of giraffe.

Other species

A few other animals sometimes interact with giraffe in the wild, including zebra and other herbivores which might appreciate the increased vigilance possibilities. Birds that commonly perch on the backs and necks of giraffe are the red-billed oxpecker (*Buphagus erythrorhynchus*) and the yellow-billed oxpecker (*B. africanus*), also called tick birds (Berry, 1973). The former were also present in the eastern Transvaal (Innis, 1958), and buffalo weavers (*Texor niger*) have been reported on giraffe in South Africa (FitzSimons, 1920).

Birds may give alarm calls if danger threatens, but oxpeckers are especially important in searching through a giraffe's hair, hunting for ticks to eat. If a giraffe has an open sore, however, the birds will also peck at these wounds, which may or may not be helpful to the giraffe. It seems likely that a bird could spread infection as it flies from one giraffe with open sores to another, or even to an animal of another species.

The two oxpeckers also interact with giraffe in Namibia, along with cattle egrets (*Ardeola ibis*) and fork-tailed drongos (*Dicrurus adsimilis*), which prey upon insects incidentally flushed up by giraffe as they move about (Fennessy, 2003). In northwestern Namibia where there are no oxpeckers, an additional gleaner on giraffe is the palewinged starling (*Onychognathus nabourup*). These birds peck away on the mane, back, underbelly and tail hair, but when they invade a giraffe's ears and head, their host shakes them off.

Man-made structures

In the wild, giraffe don't mingle with human beings if they can help it, but man-made structures beyond snares may prove deadly to them. In Marloth Park Reserve bordering Kruger National Park in South Africa, four giraffe were electrocuted and died in one year after walking into power cables hanging from poles (Anon., 2012b). (The reserve is densely wooded, so giraffe often choose more open areas for browsing where there are roads and wires.) Rangers have piled stones around the electricity poles as a barrier to giraffe, but better measures are being considered as other animals have also been electrocuted, including a martial eagle, a warthog and civets.

In earlier days, giraffe broke fences that were newly erected on their ranges, but soon they learned to step over rather than barge through them (Innis, 1958). Dagg and Foster (1982) document other disasters too: giraffe used for target practice by passing vehicles; telegraph wires hung so low that they were broken, often along with their poles, by giraffe; veld fires; and collisions that derailed railway engines, damaged cars and killed giraffe. Fortunately such accidents are rare.

Ethogram

An ethogram is an inventory of all behaviours or actions performed by a species of animal. Ideally, the behaviours are defined objectively, are mutually exclusive and include no inference about what they might mean. Peter Seeber, Isabelle Ciofolo and André Ganswindt (2012a) have now developed an ethogram for giraffe which lists 65 different behavioural patterns grouped into categories. These categories, which apply to both wild and captivity giraffe, are: general activities, abnormal repetitive behaviours, general interactions, bull–cow behaviour, bull–bull behaviour, cow–cow behaviour, maternal behaviours and interactions by calves. The terminology should prove useful in harmonizing behavioural observations in the future.

5   Individual behaviours

This chapter deals with the gaits and activities of individual giraffe with the exception of reproductive behaviour (Chapter 9) and the huge topic of feeding (Chapter 3).

Daytime activity patterns

In the early 1970s, Barbara and Walter Leuthold (1978b) documented over 230 hours what 12 giraffe did each day at different times of the year in Tsavo East National Park (their nightly activities remaining even yet a mystery). Their subjects were seven adult males, one subadult male and four adult females whose main preoccupation was usually food. The males spent from 15% to 49% of their time feeding, compared to 25–70% for the females. The males, who had more time on their hands, often necked or sparred in the mornings, as they had done also at Fleur de Lys ranch in South Africa (Innis, 1958). If giraffe chose to lie down at midday, they did so more often in large than in small groups, perhaps for security. If a male were after a female in oestrus, their feeding time, greatly reduced, was replaced by sexual activities. When giraffe weren't feeding, they were often ruminating, looking blandly about as they did so.

Individuals do a large number of other things too, as we shall see in this chapter.

Walk

I was determined to do research for my doctoral thesis in biology at the University of Waterloo on a topic that did not involve hurting animals, so I decided to compare the gaits of various large mammals (Dagg, 1979). I already had many film sequences of giraffe of various sizes walking and running that I had taken in 1956, so I set about borrowing film footage of other species as well, specifically that of okapi, American antelope, six species of bovids and nine species of deer.

To analyse a gait, I organized a projector on a box with each frame of film projecting onto a high table; I was working at home, so standing prevented my two toddlers from climbing onto my lap with their crayons to 'help' me. Then I jotted down which feet were on the ground of a walking or running animal in successive frames, many, many thousands of them. (Nowadays a computer would probably collect such data in a few days, rather than a year or more, sigh.)

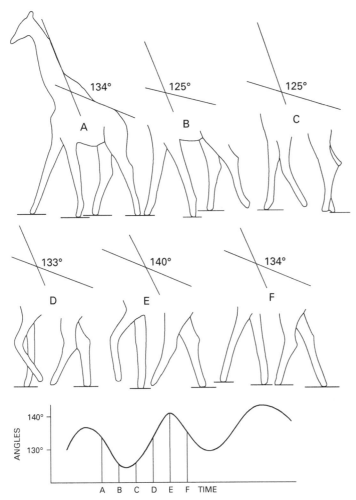

Figure 5.1   Half a walking stride of an adult giraffe showing neck–back angles at equal time intervals. After Dagg (1962b).

For the stately walk of the long-legged giraffe, soon after the left hind leg begins to swing forward the front left leg does too. When these two legs are again on the ground so that the animal is supported by all four legs, the right hind leg begins its swing, followed by the right front leg (Figure 5.1A). This lateral type of walk contrasts with diagonal walks present in the smaller non-giraffids; white-tailed deer spent about 20% of the time of a walking stride on diagonal legs (left hind and right front or right hind and left front; Figure 5.2b), whereas walking giraffe and okapi were never supported by diagonal legs (Dagg, 1960).[1] Their walk is still stable, however, because the giraffids are supported by four legs during part of each stride – 15% of the time of the stride for adult giraffe, 14%

---

[1] The diagonal type of walk of the deer is more primitive than that of the giraffe because it offers more support to the individual; amphibians and reptiles always use diagonal legs rather than lateral legs as supports.

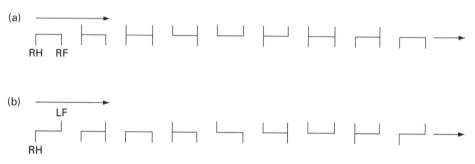

Figure 5.2   Depiction of legs that are on the ground in walking strides of giraffe (**a**) and white-tailed deer (**b**). R, right; L, left. By Dagg in Dagg and Foster (1982).

for giraffe calves and 7% for okapi (Dagg and de Vos, 1968a). The non-giraffids were rarely or never supported by all four legs during a walking stride.

This unusual lateral type of walking gait presumably evolved for giraffe because:

(a) they have long legs which would otherwise collide on either side as the hind leg swings forward first;

(b) swinging two right or two left legs forward at nearly the same time may require less muscular energy for the trunk (pacing horses which move their lateral legs in synchrony are slightly faster than trotting ones);

(c) it allows a longer stride with the hind foot set down ahead of where the front hoof had been placed (Figure 5.3);

(d) giraffe have a large mass which keeps their body relatively stable when on two lateral legs; and

(e) they live often in fairly open areas where they can spot predators from a distance. By contrast, cervids live in wooded areas where a much more diagonal gait is useful and balanced if an individual wants to stop suddenly to check out signs of danger. Camels which live in completely open areas have a lateral walking gait and pace rather than trot when moving more quickly (Dagg, 1974).

Travelling

In areas with plenty of forage, giraffe may not travel far each day. At Fleur de Lys ranch during the first three months of the growing season, giraffe moved on average only 1.1 km in 20-hour periods (Innis, 1958). However, in dry areas such as Niger where food is often sparse, giraffe are champion walkers. Yvonnick Le Pendu and Isabelle Ciofolo (1999) who studied the seasonal movements of giraffe over a 15-month period found that 19 chose to visit the Fandou region east of Niamey, a distance of 80 km which they could cover in 5 days. However, one-third of the herd went much farther afield than the others. Two travelled 160 km to the Gorou-Bassounga Forest. Others migrated a distance of 300 km, one of the largest movements ever recorded.

Why did these *G. c. peralta* giraffe cover such long distances?

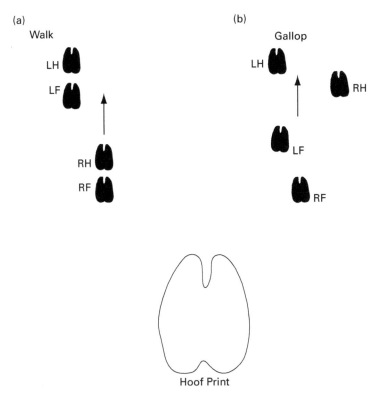

Figure 5.3   Footprints of giraffe walking (**a**) and galloping (**b**). R, right; L, left; F, front; H, hind.
By Dagg (1968). Artist: Jean Stevenson.

- It may reflect that individuals originally from Mali had been forced to move east because of poaching, deforestation and drought, but still remembered the good feeding grounds 'back home'.
- The best feeding grounds in deserts are along current or ancient rivers where there is permanent water. This means giraffe tend to follow a linear path rather than just browsing in one area, a path that can be more easily measured (although giraffe movements within one area might be just as great, even if not measurable).

Namibia is a dry country like Niger, so walking is important there too. Julian Fennessy (2009) traced the routes of giraffe with GPS satellite collars. The average daily walk of three bulls was 5.64 km, and for females 1.87 km. There was no correlation between the daily distance travelled and the mean ambient temperature, but there was between travel and the time of the walks: giraffe were active around dawn, moving and feeding; travelled little during the middle of the day when it was very hot (up to 40°C); and became active again around dusk when it was cooler (about 20 or 25°C).

Some examples of daily movements of giraffe:

- Cape herd, growing season, average 0.21 km/h during daytime (range 0–0.76 km/h) ($n = 26$) (Innis, 1958);

- Cape herd, growing season, average 1.1 km/h during 20 h overnight period (range 0–2.6 km) ($n = 23$) (Innis 1958);
- Cape males average 5.9 km/day; females average 2.9 km/day (Langman, 1973);
- Cape giraffe – 8 collared individuals; average 5.3 km/day (Deacon, 2012);
- Zambia males average 3 km/day; females average about 2.3 km/day. (One male travelled 140 km (Berry, 1978). Most of the adults were known to have travelled at least 40 km);
- Garamba National Park giraffe with similar results to those of Innis (Backhaus, 1961); (in the Kruger National Park one female walked 20 km and a male covered 60 km in a straight line in 2 weeks (du Toit, 1990a).)

Home ranges

In Niger, as for other giraffe living near people, home ranges are affected by poaching, fences, agricultural fields and livestock as well as by topography, rainfall, vegetation and the presence of water (Suraud, 2011). Giraffe are unique in Niger in that most have a large home range which enables them to take advantage of regions where and when their preferred vegetation is available. In the wet season on the Koure plateau, highly nutritious browse plants provide ample food for giraffe. Most then move in the dry season to the Harikanassou area where there is permanent water to drink and edible browse along with farm fields and fallow lands. Farmers grow crops in these regions when there is rain while herders are perpetually on the move tending cows, oxen, goats, sheep and camels. Two-thirds of the giraffe population, both males and females, spend their whole year in the core, moving to the Koure plateau for the rainy season and then back to the Harikanassou area. The mean size of home ranges of these resident giraffe was 47 km^2 in the rainy season and 91 km^2 in the dry season.

Home ranges of giraffe vary primarily with availability of food, but also with other factors including seasonal or variable rainfall, presence of water, temperature, number of giraffe in a group, composition of each group, number of other herbivores and predator density. The following are examples of home ranges.

- In Nairobi National Park, for commonly seen individuals, males had an average home range of 62 km^2 and females of 85 km^2 (Foster, 1966).
- In Timbivati Private Nature reserve, South Africa, home ranges were 22.8 km^2 for males, 24.6 km^2 for females and 12.8 km^2 for juveniles (Langman, 1973b).
- In the Luangwa River area, Zambia, the largest male had a range of 145 km^2, and the largest female, 82 km^2 (Berry, 1978). The average home range for 14 males was 82 km^2, and for 6 females 68 km^2.
- In Tsavo East National Park, males and females had an average home range of 160 km^2. The maximum home range for a male was 650 km^2 and for a female 480 km^2 (Leuthold and Leuthold, 1978a). The home ranges of the males were more variable, with young males moving more on average than did females, and old males becoming more sedentary– nine of them had home ranges smaller than 20 km^2.

- In Lake Manyara National Park where there was heavy vegetation, females and males had home range averages of 8.6 km^2 (although there was a large variation among individuals, some of whom had poor body condition) (van der Jeugd and Prins, 2000).
- Arusha National Park in 1980 had an area of 119 km^2 with dense vegetation and a population of 471 giraffe (along with many other prey species but few predators) (Pratt and Anderson, 1982).
- In northwestern Namibia, Fennessy (2009) worked out the home ranges of three giraffe whom he had collared. Two bulls had the largest ever recorded for any giraffe in Africa – 1950 and 1627 km^2. The home range of the cow giraffe, 1098 km^2, was the second largest for a female. There were no regular seasonal movements or seasonal home ranges as there were in Niger. Strangely, despite the arid environment of the northern Namib Desert, the presence of surface water did not seem to influence giraffe movements; fewer than 10 sightings in 70 years of giraffe drinking water have been recorded.

Gallop

Because of their long legs giraffe are unable to trot like other ungulates, so their only fast gait is a gallop in which the legs move asymmetrically and variably (Dagg and de Vos, 1968b). Because the gait is fast and the legs long, it is very difficult to film. Ideally, one would stand still with a movie camera while a giraffe galloped in an arc around one, keeping the same distance away, but this would require a miracle.

The power and weight of the giraffe are more in the forequarters than the hindquarters, so the main propulsion of each stride comes from the forelegs. The gallop can be either slow or fast – as speedy as 56 km/h (35 mph). However, because of their smaller mass, cheetah, pronghorn and gazelles are far faster, at up to 100 km/h. The giraffe has less need for speed, as it is too large to be attacked by any predators other than lions, from which it usually defends itself with kicks. When a herd is flushed perhaps by human hunters, however, the smaller giraffe run faster than their elders. Both have periods in a galloping stride when the body is supported by a single foot, by two front feet and by two hind feet (Figure 5.4), but the smaller giraffe in this sequence has two periods of suspension when no legs are on the ground (once while its legs are bunched together and once when they are extended), while the larger animal has no such period (Figure 5.5). The faster an animal runs, the less time its feet are on the ground. When they gallop, except when they are changing leads so that the major impetus switches from one side to the other, giraffe set their feet down in a rotary manner – left hind, right hind, right front left front, or vice versa.

Role of the neck in giraffe movements

As a giraffe walks in its stately manner, its neck swings forward along with two lateral legs, shifting the centre of gravity ahead (Dagg, 1962b). It is straightest when the two legs on one side of the body are beginning their swing forward, and most backward leaning

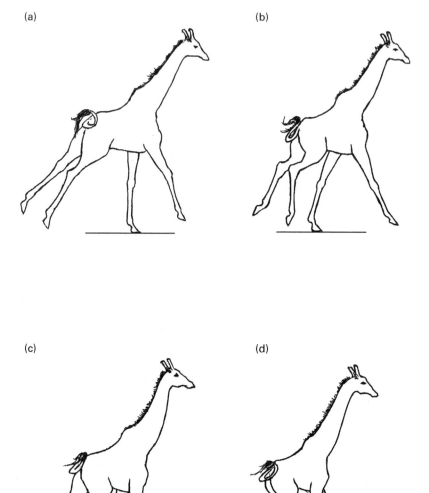

Figure 5.4   Gallop of a subadult giraffe: **a**, left foreleg newly landed; **b**, left foreleg about to push off;
c, right foreleg newly landed; **d**, all feet in the air, with right left hind leg about to land. By Dagg
in Dagg and Foster (1982). Artist: Jean Stevenson.

when these legs are about to be set down (Figure 5.1). Therefore, there are two swings of
the neck per stride which accompany the forward movements of the left legs and then the
right legs like a pendulum. The neck movement of young calves is less extreme; they
walk with their neck more erect, presumably because their neck is relatively short (1/6 of its
total height) than that of the adult giraffe (3/8 or 1/3 of its total height). While galloping,
giraffe also move their necks back and forth, one pendulum swing per stride. The neck of
a young giraffe is most stretched forward when the animal is in the air, midway between

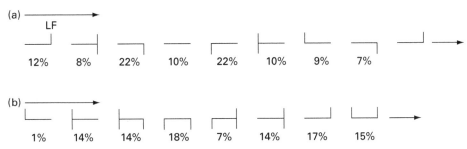

Figure 5.5   Depiction of legs that are on the ground in galloping strides of a young giraffe (**a**) and of an adult male giraffe (**b**), with percentage of stride in each phase. By Dagg in Dagg and Foster (1982).

having launched the body and its landing; when a foot hits the ground the neck is drawing back to stabilize the movement.

When a giraffe at full gallop makes a sharp turn, it uses its neck as a counterbalance, shifting it in the opposite direction to the turn, just as a person on a fast motorcycle shifts his or her weight left if turning right (see photo, Anon., 2012h).

Lying down and getting up

For giraffe, lying down is easier than getting up. They fold their front legs under them first, then bend their hind legs to settle down on their sternum with a hind leg sticking out either to the left or the right of their torso. None lies squarely on its hindquarters as a camel does. Young giraffe lie down more often than do older ones. Vaughan Langman (1977) found that Cape giraffe less than 2 months of age lay down 79% of the day on average, while calves 2–18 months lay down 19% of each day. In Tsavo National Park, Kenya, the Leutholds (1978b) did not see females lying down, but male giraffe did so for on average 13% of each day. Lying down in zoos takes longer than in the wild, up to 15 s, because the giraffe first carefully checks out the flooring and the placement of its spread hind feet and folding front feet (Kristal and Noonan, 1979b). Some old giraffe never do lie down (Dagg, 1970b).

Because of their large mass, giraffe may have difficulty getting up after lying down (Figure 5.6). Their neck movements are vital in accomplishing this (Dagg, 1962b). From a resting position first a giraffe throws its neck back for impetus so that it can hoist its forequarters onto its 'foreknees'. Next it moves its neck forward again for balance, pauses, then moves its neck backward slightly before throwing it forward so this new impetus allows the animal to shift its hindquarters onto its back feet. It holds this pose for nearly a second, then draws its neck back again along with the weight of its body so that it can change from a kneeling position of its front feet to standing on them. This sequence lasts about 4 s in an agile subadult.

Surmounting fences

When the low veld of northeast South Africa was carved up into cattle ranches following World War II, the giraffe that had been roaming free there for millennia were

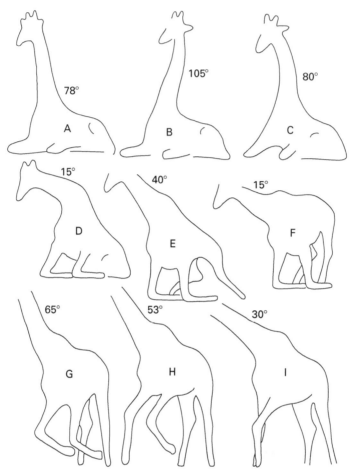

Figure 5.6   Sequences of a recumbent adult giraffe getting to its feet, giving angles of the neck to the ground. From Dagg (1962b).

hemmed in. At first, they broke through so many of these wire 4.5-feet high fences that the ranchers worried they would have to shoot them. Soon, however, they learned to step over them (Figure 5.7). To do so they first look down carefully at the barrier, then pull their neck back and bodies up until they are standing very tall on their hind legs. This enables them to step over the top wire with their two front legs. Then they shift forward with the fence under their belly before putting all their weight on their front legs so they can hop their hind legs over too. Giraffe no longer broke the fences in order to cross them. However, this ability apparently led to one disaster I witnessed: a mother giraffe stepped over a wire fence into a different paddock, then waited for her newborn calf to follow her. This was impossible for it. The mother waited about for a while, then moved off on her own, leaving her baby behind. The next morning there was no sign of either animal, but the youngster was thought to have perished.

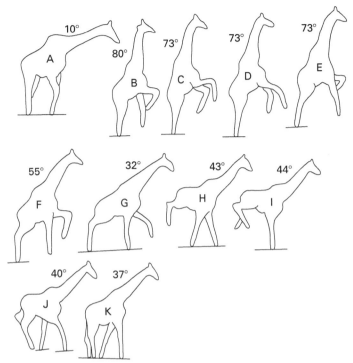

Figure 5.7   Sequences of an adult bull jumping a 1.5-m fence, giving angles of the neck to the ground. From Dagg (1962b).

Bending to the ground

It's a long way down for a giraffe to drink, lick at salt on the ground or pick up a fallen branch, but they manage well despite the rapid change in blood pressure for the brain (see Chapter 8). Captive individuals tend to bend down in the same way each time, with forelegs bent, or forelegs straight and therefore apart, or forelegs straight and slightly apart leaning over one shoulder (Dagg and Foster, 1982, p. 106). To attain this last awkward stance, the far hind leg usually leaves the ground. If the flooring is smooth at a zoo, giraffe will not spread their front legs far apart for fear of slipping.

Drinking

It is impossible to know how giraffe feel about water; some drink lots of it and others seldom drink at all (Foster and Dagg, 1972). Observers have reported that when water is readily available, giraffe drink often, even daily and at any hour of the day or night. They drink at rivers, dams, pools and even cattle troughs (which other species are often too nervous to approach). At zoos they drink especially frequently because they do not have much or any browse high in water content; at an ambient temperature of 32°C, one captive giraffe drank about 45 litres a day (Cully, 1958). Various observers have written that giraffe with whom they were familiar seldom or never drank water, even in the dry areas of the Namib Desert (Fennessy, 2009). There is general agreement that giraffe are physiologically adapted to withstand droughts.

There is a huge advantage in being able to survive on the water supplied by vegetation. It enables giraffe to live in semi-desert areas too dry for many other species. It also means that they need to spend little or no time drinking in the wild which is when they are most vulnerable to lion attack, and where the ground may be too marshy for giraffe to stand firmly.

Giraffe reach down to drink in the wild either by straddling their front legs or by bending them instead at the 'knees'. They often drink from 20 s to a minute without pausing before they straighten up again. It is an effort to bend down and struggle up, but I saw giraffe do this as many as six times during one trip to the water. Perhaps this facilitates swallowing, or the drinking stance is too tiring to hold for long, or they want to check for possible danger.

Small calves do not drink water at all. Maybe their legs are too long or their neck too short for their mouth to reach the water? They obtain moisture instead from their mother's milk and from a few nibbled leaves, but that is all. Vaughan Langman (1982) believes that this is why a youngster hides among vegetation in the shade during its first few months when its mother goes off to forage for herself. If her baby followed her, it might overheat and suffer dehydration because, like its parents, it can't pant or sweat. It does raise its body temperature when it is hot, as adult giraffe do, but it is not large enough yet in body size to cope with great heat as do adults.

Before I saw giraffe drinking, I had assumed they would lower and lift their heads slowly because of the rapid change in blood pressure at their brains. Not so. When I filmed this activity, I found they whipped up their heads to a horizontal position in half a second, then rested there briefly while shifting their front legs under them. Then they whisked their heads up to about 45° above horizontal in another half second to second. My presumption was obviously completely wrong. (See Chapter 8 for a discussion of blood pressure changes.)

A study is now underway in the Ongava Game Reserve in northern Namibia adjacent to Etosha National Park in which Ken and Sabina Stratford (2011) are using camera traps at waterholes where giraffe drink to collect information: how often do giraffe drink at each station; is there a difference between males, females and young; what waterholes are preferred; what is their seasonal use? Already they have found:

(a) that giraffe drink at different hours near human habitations: peak 23:00 h there compared to peak 13:00–15:00 h at remote waterholes;
(b) that males visit more sites to drink than do females; and
(c) that in semi-arid areas giraffe do not drink at waterholes in regular patterns.

Rubbing and scratching

In the eastern low veld of South Africa, giraffe spent much time rubbing and scratching themselves despite their thick skin (Innis, 1958); ticks were common on their underparts where there is little hair, especially in the rainy season. Giraffe can't shake their skin as horses do to dislodge an insect or bird because their skin is too tightly attached to their

Figure 5.8   Rothschild's female drinking with splayed legs. Photo by Zoe Muller.

Figure 5.9   Male Rothschild's giraffe scratching his hind leg. Photo by Zoe Muller.

torso. If one wanted to scratch its stomach, it might straddle a 6-foot bush and rock back and forth over it. For the torso, he or she might push into a dense shrub. For an itchy head, a giraffe rubbed it on a tree, in bushes, on the head or neck of a friend or even on the ground. A foot would be dragged back and forth among branches. Giraffe in northwestern Namibia where the tick load is high also rub themselves against small shrubs and bushes to remove parasites such as large ticks (Fennessy, 2003).

With its teeth or tongue an adult giraffe can nibble or lick a local body part that it can reach, but this does not include the lower neck, although a baby giraffe can reach this area. Nor, because of their shape, can they scratch their heads with their hind hooves as most ungulates can. Dieter Backhaus (1961), who watched a newborn giraffe unsuccessfully trying to do this, cites it as an example of evolution: more usual-shaped animals could presumably touch their head to their foot such as giraffe ancestors, but the long neck of the giraffe prevents this.

Body postures

When they are walking or standing, large giraffe hold their necks at about a 50 or 60° angle above the horizontal, while the necks of babies are rather straighter, at an angle of perhaps 70°. When a giraffe shakes its head, it holds its neck to about 45° from the horizontal. A giraffe may intimidate another giraffe or a lion by lowering its neck until it is horizontal with the ground. If danger threatens, it lifts its head high, looking about, its mouth pointing to the earth and its nostrils flaring. When preparing to flee, it turns its body at an angle to the threat. If a male is involved in a fight with another male, he may hold his head high at an angle of more than 180° with his neck in a dominance position. When irritated or ready to kick out, a giraffe may paw the ground with a forefoot.

Rumination

Giraffe in the wild spend many hours a day browsing, and many more ruminating. After leaves, fruits, twigs or flowers picked from trees or bushes are swallowed by a giraffe, they descend to the large rumen where they are mixed with gastric juices and partly fermented. While the animal is resting often at midday after a morning of browsing, softened food from the rumen that has been formed into a bolus shoots up the muscular oesophagus by peristaltic action into the mouth, where the food is masticated more thoroughly than before, with the lower jaw acting in a circular motion as it pushes the bolus against the hard palate at the roof of the mouth. The rate of chewing the bolus is about one chew per second and, depending on its size and composition, it can receive from 29 to 81 chews (Backhaus, 1961). When the bolus is swallowed, this time the food bypasses the rumen and proceeds to the large abomasum or psalterium. Three or four seconds after this, another ascends to the mouth to be chewed in turn (see Chapter 8 with regards to digestion). At Fleur de Lys ranch in South Africa, the time spent ruminating increased as the cold season gave way to summer. In Kenya, Masai giraffe ruminated for about 20% of each day, the youngest beginning this activity briefly at about four weeks of age (Leuthold, 1979). By contrast, Vaughan Langman (1977) noted that rumination began in young 4–6 months old.

Presumably rumination evolved so that large browsing or grazing animals of the suborder Ruminantia (of the Order Artiodactyla) could feed quickly in unprotected areas where they were vulnerable to attack by predators, then retreat to safer areas where they could rechew food from their rumen at their leisure, looking around at the same time to watch for predators. It is a relaxing activity which may occur throughout the day or night, in long bouts lasting up to 75 minutes or in short spurts (Innis, 1958; Wyatt, 1969). Giraffe chew their cud while standing, walking or lying down, but not in deep sleep.

Rest

On the Fleur de Lys ranch in South Africa in the heat of midday, all giraffe activity came to a stop. After a morning of hearty feeding, the animals seemed to have lost their starch. They often stood or lay about on an open slope, 20 m apart from each other, often not even chewing their cud. They faced any which way, almost in a trance, their necks and heads drooping, their eyelids almost closed, their tails still. (Indeed, tail movements indicate energy, the opposite of rest; they twist or thrash when giraffe run or when one is involved in fighting.) Rest time would have been an ideal chance for a lion attack, but these predators were surely comatose as well. I had to rub my eyes myself to stay awake, watching them from my small hot car.

Giraffe often rest while standing, but they certainly do so while lying down. This non-activity is most common in young giraffe, supposedly because, for adults, it requires a great deal of effort for older animals both to fold up their long legs to get down, then later reorganize them to get up again. In the Taronga Zoo I almost never saw giraffe older than 5 years lying down (Dagg, 1970b).

Sleep

Large ungulates have little deep sleep each night. By contrast, predators and small animals sleep soundly for much longer periods. Rodents and shrews snooze snug in their burrows while predators can relax because they are the hunters, not the hunted (Kristal and Noonan, 1979b). Based on these generalities, sleep length could be correlated with large body mass, large brain mass or slow metabolic rate, but 'danger' is certainly a good predictor of limited sleep time for ungulates.

Like other ungulates, giraffe are not great sleepers, having only small amounts of fragmented shut-eye each night, and perhaps some dozing during the daytime. It is difficult even to know when giraffe are sleeping because, judging from EEG recordings of domestic herbivores, they may do so when standing as well as lying. Presumably the need in the wild to forage and ruminate as well as be alert for predators is why giraffe sleep so little.

To study sleep in this species, Irene Tobler and B. Schwierin (1996) made continuous video recordings over 152 nights of giraffe in a Dutch zoo, five adults, two immatures and one juvenile. During each night the giraffe slept for varying amounts of time in three postures: standing up (SS), lying down with neck erect for brief periods (RS for recumbent sleep), or lying down with the head resting on their flank (in youngsters) or the ground behind them for much shorter sessions (DS for deep sleep or paradoxical sleep). In the past only deep sleep was considered as real sleep; the other postures were labelled 'dozing'. The two lying-down positions also occur in wild giraffe, but may not be as common because if danger threatens, it takes several seconds for an animal to get to its feet so it can flee.

In standing sleep an animal remains motionless, not moving even its ears, its neck drooping. Its sleep is often broken by short waking intervals when it raises its neck to look around and moves its ears before sinking back into slumber. In a recumbent position a giraffe lies on its brisket and abdomen or flank with its legs folded beside it and its neck erect, while often chewing its cud. During deep sleep each animal is completely relaxed with its neck and head stretched back along its side (although sometimes their neck or ears twitch) and it does not ruminate; this deepest sleep stage lasts for only a few minutes at a time. Calves sleep lying down for long periods after birth, while old giraffe spend more time sleeping on their feet than do younger animals, probably because it is more difficult for them physically to lie down and get up.

In their barn at night the giraffe all slept in a bimodal pattern between about 20:00 and 07:00 h, more deeply at first with a small trough of wakefulness between 2:00 and 4:00 h, and then less soundly toward dawn. In the daytime they sometimes napped for brief periods. Even though Tobler and Schwierin broadened the definition of 'sleep', it was still short-lived and episodic. Total sleep time (TST) for adults averaged 4.6 h during each 24-h day, with DS only 4.7% of this – most RS episodes lasted less than 11 min.

The amount of sleep for the giraffe is internally regulated:

- the TST was similar for the adults, but the times for RS (usually more for the young) and SS (usually more for the oldest) varied;
- if the giraffe had spent some hours outside rather than inside during the day, the first peak of their sleep profile was absent and they had more RS time in the second half of the night; and
- if a giraffe napped during the daytime, it had slightly less RS and an increase in SS the following night, perhaps indicating that SS is a 'lighter' sleep than RS.

Florian Sicks (2012) at the Berlin Zoo chose to study REM-sleep positions in giraffe for his PhD thesis. REM (rapid eye movement) in giraffe means resting their head on their flank or on the ground beside their hind legs, as in deep sleep defined above. He had the unusual job of watching 17 giraffe sleep for a total of 9675 h, wondering if their awkward sleeping positions could be correlated with stress in their lives such as lack of food, changes in their group mates, giving birth or being transported. He checked stress levels by measuring cortisol metabolite in their faeces via enzyme immunoassay, the first time this had been done for giraffe. Sicks found that normally all the giraffe slept on average 27 min at night in the REM-sleep position, with juveniles spending 63 min in this position compared to only 4.5 min for giraffe over 20 years of age.

When the giraffe were under stress, however, their sleep patterns changed:

- after being transported all (four) giraffe had no or almost no REM-sleep during the first nights, along with an increase in cortisol metabolite concentration in their faeces for several days;
- after the death of the adult bull, a female had no REM-sleep for 21 days, although her cortisol metabolite concentrations did not change; and
- newborn giraffe that suffered malnutrition and died after some days spent significantly more time in REM-sleep than did healthy newborn giraffe.

Sicks concludes that REM-sleep which can be assessed by video is well suited as an indicator of stress in giraffe and can be used to monitor their well-being.

In research carried out on two female adults at the Buffalo Zoo, one heavily pregnant, Mark Kristal and Michael Noonan (1979a) report that the breathing rate while lying down at rest was about 15–20 breaths/min and the pulse rate about 50 beats/min. When the baby was born, on its first two days of life it spent about 25% of the total time sleeping, and 90% of that time in the deep sleep posture with its head on its flank.

Figure 5.10   Subadult Rothschild's males sparring with an onlooker. Photo by Zoe Muller.

For Cape giraffe in the wild, Langman (1977) found that RS was greatest in young calves (0–60 days), and least in immature animals (1.5–3 years).

Necking, sparring and mounting of males

Where subadult and adult males make up a large proportion of a giraffe population, such as was the case at Fleur de Lys ranch in South Africa where I observed wild giraffe in 1956–1957, I often saw males rubbing their bodies together, necking, sparring or one male mounting another (Innis, 1958).[2] These occurred during all hours of the day and in

[2] I was apparently the first person to document homosexual behaviour in wild animals; it must have occurred in zoos, but presumably was thought to be caused there by unnatural captive conditions. At Fleur de Lys ranch it was much more common than mating behaviour, which I only saw once. I was too embarrassed to mention it to anyone, even those at the ranch, but did describe it in my final report (Innis, 1958). Many people requested information from me related to this, including Julian Huxley, the renowned zoologist.

   When my late aunt read about homosexual behaviour in my paper, she was horrified. I heard her telling my uncle, 'They shouldn't let young girls see such things.' I have since often laughed to myself at the thought of officers of some sort lurking about where young women are observing wild animals. Should they distract one if she were about to see something untoward? Or spook giraffe who were being licentious? And who would pay for this unusual posse?

   Much later when I worked at a university where lesbians were being abused for their 'unnatural' proclivities, I reviewed the behaviour of scores of animal species in the wild to show that most of them

Figure 5.11   One fighting reticulated male has his opponent's hind leg caught over his neck. Photo by Karl Ammann.

every season. Males were little involved with females unless they were hoping to mate or mating, which took far less time.

Necking involved one male gently rubbing his head or neck against the body of another, or the two males mutually rubbing their trunks and necks together. This often sexually aroused one so that he mounted the other or even several other males. Sometimes after one male had mounted another they might begin to neck, then spar, and then fight fiercely with heavy blows from the head taking the place of gentle rubbing. Then they might stop fighting, and one might again mount the other. Once a male with erect penis was about to mount another when that male walked away instead. With his penis still erect, the male moved to another bull to mount him. One male with an erection rubbed his penis back and forth among the branches of a bush; while he did this his tail stuck straight out and then curled tightly over his back. There seemed to be no jealousy among partners. Such mounting behaviour took place throughout the year, and when females were nearby. The pink penis is about 8 inches long when it comes out of its sheath, with a downward hook at the end, rather like a crochet hook. I sometimes saw liquid dripping from the tip.

practised homosexual behaviour and that therefore it was not perverse, but perfectly natural (Dagg, 1984). This pioneering effort was much surpassed later by the momentous work of Bruce Bagemihl (1999).

Recently it has been discovered that Murray Levick with the 1910–1913 Scott Antarctic Expedition did see Adelie penguins engaged in homosexual behaviour but, torn between prurience and scientific accuracy, he documented his observations in Greek so that only educated gentlemen would know 'the horrors he had witnessed' (*Toronto Globe and Mail* 2012, 11 June, A16).

Sparring occurred among either two or three males of varying ages. Two might stand side by side, facing the same or the opposite direction, swinging their head at each other's body with horns foremost. One may lower his head and then swing it toward the other, who dodges slightly to avoid contact. The horns may miss or glance off him. Blows are seldom heavy, with the two animals between bouts of hitting standing and rubbing their necks together or just looking off into the distance. Once a calf sparred between two large bulls, each tapping him gently from each side, the baby waving his head about at all sides, sometimes missing any target completely – sparring first appeared in male calves only a month old (Leuthold, 1979). The necks swing back and forth gracefully during such sparring matches but there is little foot movement.

In serious combat there is a great deal of leg movement as the bulls advance and retreat between heavy blows, but never kicks which are reserved for predators. A match may develop from sparring, or it may be a direct challenge for dominance (Coe, 1967). If the latter, one large male approaches another while assuming a 'proud' posture with his shoulders directed toward his opponent and his legs and erect neck held rigid. If his opponent also assumes a similar 'proud' posture, the two place themselves side by side, and head to tail, the tail itself often twisting about. They each straddle their legs to obtain a firm stance, then curve their necks away in preparation for the first blows. They take turns swinging their heads at some body part of the other, their own legs always spread for stability. After one lands a body blow with a loud thwump, the other may retaliate with a hit to the legs or neck. One winning male nearly upended his opponent whose leg got caught on his head and was then straddled over his neck. Between blows they may stand motionless as if resting, or march a short distance side by side, their necks stretched parallel to the ground or extended high in the air. I saw one pair circling a tree at least 10 times while they pounded each other's flanks. Two contestants at Fleur de Lys were so involved in their battle that they ignored an African fence repairman walking past within 10 m of them. A fight may degenerate into sparring but sometimes it is more serious, with one of the contenders galloping off followed by the other. Or it may end with one bloodied and knocked unconscious. One loser died because his top neck vertebra had been splintered by a blow, with a sliver piercing his spinal cord (Clerck, 1965).

A new behaviour involved in a contest called 'leg lifting' has been reported by Dale Peterson and documented on film by Karl Ammann for both Masai and reticulated giraffe (Anon., 2012e). Here, during a match, one male may lift his front leg at the other, or a giraffe's leg may be caught on the other's back or neck. During sparring, one male may repeatedly put his neck and head under a leg of his opponent in an apparent effort to destabilize him (pers. obs.).

In Kenya, Warden Jamie Gaymer (2011) was moving reticulated giraffe from one reserve to a new area to alleviate pressure on the browse. This meant that breeding males unfamiliar with each other were coming into contact, because he watched a vicious fight between two of them. One bull was knocked off his feet by a head blow to his chest from the other. The winner tried to hit the downed male again, but was chased off by the human observers. The loser who struggled to his feet after a few minutes seemed to have big hematomas on his shoulders where he had been whacked.

In a similar scenario in Tanzania's Lake Manyara National Park where there was a stable core of females, the local large males never fought or necked among themselves,

Figure 5.12   Male in unfortunate position with his opponent's leg stuck between his horns. Note hoof anatomy and faecal pellets. Photo by Karl Ammann.

but when an immigrant male entered the area he was always challenged by a resident male not actually to a fight, but rather to 'high-intensity necking' (van der Jeugd and Prins, 2000). This was thought to be a temporary intrusion by the newcomers into a dry season refuge being rebuffed by a local resident.

Serious combat determines dominance, giving the winner precedence at mating with females in oestrus. This implies that males know each other individually even though they are often loners and seldom together in the same group (Foster and Dagg, 1972). Malcolm Coe (1967) believed that necking and sparring was a 'sexuo-social bonding mechanism' for males which fostered some cohesion within a group, perhaps similar to that present in mountain sheep (Geist, 1975), although this possibility applies only to young adult male giraffe. This means that male giraffe in an area are familiar with each other. Decades later we now know that giraffe can communicate with infrasound, which makes this possibility highly likely.

Head-hitting battles between male giraffe may seem ferocious and have caused death, but they seldom cause serious injury. The tops of their horns are blunt and smooth, and blows fall mainly on the trunk and neck where the skin is thick. By contrast, when giraffe are in danger from lions, they respond not with head blows but with kicks from either front or hind legs. In one YouTube sequence, a female with a calf seems to have killed one lion with such fierce blows and she is effectively fending off the other pride members.

Figure 5.13   One of the first published photographs of the leg lift behaviour in Masai males. Photo by Karl Ammann.

Figure 5.14   Winner of a fight chasing away the loser. Photo by Zoe Muller.

Vigilance

Theoretically, species preyed upon by carnivores should be alert at all times to possible danger. During one year, Elissa Cameron and Johan du Toit (2005) studied vigilance behaviour in giraffe living in Kruger National Park (along with lions and hyenas) by driving along tracks until they encountered a herd, which they defined as all animals within 100 m of each other, and stopped where there were most animals. This usually meant all the giraffe that they could see among the woody vegetation of *Acacia nigrescens* and *Sclerocarya birrea* trees. They recorded each individual's age (from size) and sex, then chose one adult or subadult giraffe who was feeding to be their focal subject for 20 min. This involved noting the angle of its neck at which it foraged and the number of bites per minute at each angle it positioned itself while foraging, as well as any drinking, interacting or grooming. A foraging episode was defined as all continuous bites at one of four neck angles (at 45°, 90°, 135° and 180° to the front leg). If an animal raised its head to a neck angle of 135° or 180° and stopped chewing for 1 s, this was counted as a vigilance scan. Because groups were fluid, individuals were rarely seen in the same group more than once.

A common understanding is that the larger the number of animals in a group, the safer the members are, given the increase of watchful eyes, so the results for these giraffe were surprising:

- there was little relationship between group size and vigilance, despite this correlation in many other species;
- time spent scanning by lone cows did not differ from that by cows in groups of any type;
- time spent scanning by lone bulls was less than that by bulls in groups; and
- the presence of calves did not influence the number of scans by their seniors, either male or female.

The researchers conclude that predation risk did not seem to be a modifier of vigilance scanning (although they did believe that there was some sort of vigilance behaviour in effect). However, social factors certainly did affect vigilant behaviour. For example, bulls scanned most when they were in groups with larger bulls, and least when they were with smaller bulls, while the nearness of other giraffe often affected the vigilant behaviour of both males and females.

Cameron and du Toit report that giraffe cows scanned more often than did bulls, but that bulls scanned for longer periods, so there was an equal scan effort. Scanning was far more correlated with other giraffe than with possible lions. For example, cows didn't scan more when they had a calf at foot, even though the predation of calves is high in the region, but they were significantly more vigilant when an adult bull was close. Similarly one would expect that male giraffe should scan more when alone if they were anxious about predators, but they actually scanned more when in a bull group, scoping out other males. As the largest male is usually dominant and because females come into oestrus at any time of the year, the males are presumably always conscious of possible rivals. Scanning most when larger bulls are present may help them avoid aggression from these males, while not scanning among smaller bulls indicates indifference to them. When feeding at low levels giraffe can't see much around them, nor can they when feeding high as their eyes are looking skyward. Giraffe can be most vigilant when feeding at a median height (although this is somewhat controversial). This study indicates that giraffe in this group were not much worried about lions, but interested mainly in each other.

Giraffe must be vigilant toward people, too. The number of giraffe in Ethiopia has always been small, both because of the dry environment and recently because of the growing human population, now over 77 million people. Ethiopia has declared 16% of the country as protected for wildlife, but even here it is not safe from human activity. In 2006, when there were estimated to be about 28 giraffe in the Omo National Park in southwest Ethiopia near Kenya, Pascal Fust (2009) investigated the influence of human presence on these animals. His human-impact factor (HIF) of the entire area of the Park that was potential giraffe habitat was based on:

(a) the intensity of the impact on giraffe; for example, if it was intermittent as vehicles or nomadic herders, or permanent as camps, villages or field crops; and

(b) how abundant such activity was within a radius of 2.5 km of the giraffe.

How can one assess the stress that a giraffe might feel because of such human activity? The ideal solution would be to measure its cardiac response and its blood level of glucocorticoids, but these of course could not be taken without capturing an individual and sending its stress level sky high. Fust decided instead to record the vigilant behaviour of the Omo giraffe as a percentage of their total activity budget. To this end, individual giraffe which had not been habituated to human presence were observed for sessions of 30 min at a time, during which their activity, including all vigilance behaviour, was noted down every minute. To minimize the presence of the observer's car, recordings were not begun until at least 30 min after first contact with an animal.

Fust found that one group of giraffe inhabited only about 20% of the total potential habitat of the Park, and that there was no overlap between where the giraffe lived and where people were active; this was similar to findings in Namibia (Leggett *et al.*, 2004), but not those from Niger. The vigilance behaviour of individuals in Omo National Park was all over the map, from 0% of activity budget to 93%, with most records between 0% and 35% and a mean of 25%. This figure was similar to behaviour in the Laikipia district of Kenya, where giraffe were vigilant about 21% of their time while otherwise foraging or travelling (Shorrocks, 2009). The Omo giraffe were less vigilant when there was a slight wind, but more vigilant with higher temperatures and denser cloud cover. However, there was no way to document how much vigilance might be attributed to human presence.

Fust recommends that at Omo where giraffe are likely at risk:

(a) human disturbance especially be avoided during afternoon and evenings when nearby giraffe forage extensively in the cooler temperatures;
(b) because rarely giraffe and livestock may occupy the same habitat, the use of water resources and danger of disease transmission should be studied; and
(c) further research is vital to collect more information.

Where giraffe live in extreme conditions such as deserts, where both people and predators are few, vigilance is far less. In Namibia, giraffe were vigilant at most 2% of the time (Fennessy, 2004).

Although giraffe can blend in with vegetation and be difficult to detect, even as they are vigilant, they may act in a way that ensures other animals will notice them. They are vulnerable to attack by lions and poachers, but they seem not to worry about showing themselves, even though their spotting would seem an excellent device for camouflage. When I approached a giraffe in the bush veld by car, individuals hidden among bushes and trees tended to poke their heads up or sideways so they could see me better. Otherwise I would have driven past without noticing their legs among scattered thin tree trunks. In addition, they often thrashed their tails about because of stress, again an action that drew attention to them, or they walked slowly forward while staring at a potential source of danger. Perhaps lions will be less likely to attack animals obviously aware of their presence?

Vocalizations

Giraffe can make vocal noises, but they seldom do. Once, while sitting quietly under a tree in South Africa, six giraffe wandered slowly toward me while feeding. When one spotted me she stood still, staring fixedly at me. Gradually the others, one by one, noted her posture and then my presence too. After 5 min all were gazing at me without moving. Then one snorted, as if to startle me. I remained still, thrilled. After a while they wandered away.

Giraffe occasionally moan, bleat, moo, low, cough, sneeze or, if annoyed, they may grunt, growl or snort. But they seem not to be communicating with each other. If one giraffe senses real danger, it rushes off without giving a vocal warning while the others stampede after it. Dagg and Foster (1982) cite vocal noises and their contexts reported by 30 other authors.

Sometimes giraffe make loud noises though. On YouTube there is a video of a newborn calf at a zoo being restrained while keepers examine it who gives out several loud bellows of annoyance; a staff member reports that young giraffe may vocalize but they do so less often as they grow older. Elsewhere, a mother giraffe watching such an examination of her newborn from a separate enclosure nearby was so distressed that she repeatedly gave a loud roaring sound, rather like what a dragon might make (Aim'ee Fannon Nelson, pers. comm., March 2012).

Why giraffe seldom make vocal noises we can hear seems not to stem from the laryngeal nerves (which have the left far longer than the right; Harrison, 1981), nor from the well-developed larynx which bears some resemblance to that of the horse or of herbivores able to swallow semi-liquid food while continuing to breathe (Harrison, 1980). The giraffe larynx contains flat thyro-arytaenoid muscle folds and cartilage arytaenoids, all of which can make a noise when vibrated. The difficulty in vocalizing seems rather to lie in the inability of giraffe to produce an air flow of sufficient velocity to vibrate the thyro-arytaenoid folds or arytaenoids. Perhaps such a noise is caused by a Helmholtz resonance wherein a large enclosed volume of air (as in the trachea) is coupled to outside free air by means of an aperture, in this case the nose or mouth, like a person blowing air across the top of a bottle.

We do not know if giraffe communicate with each other because we cannot even hear the infrasound noises they are making. However, recent research indicates that giraffe presumably keep in touch with others by using infrasounds so low that the human ear cannot detect them. Human beings can hear noises down to about 16 cycles/s, but giraffe and other animals such as elephants, rhinoceros and okapi, can hear much lower sounds. E. von Muggenthaler and four colleagues (2001) made recordings of 11 reticulate giraffe at zoos in North and South Carolina (see http://www.animalvoice.com/giraffe.htm). They found that the dominant frequencies used by the animals were between 20 and 40 Hz. (A hertz is a frequency measurement of one cycle per second.) A behaviour described as a 'neck stretch' (in which the head and neck start at chest level and then are thrown back over the body and curled upwards until the nose is straight up in the air) appeared to be correlated with these signals, indicating that again Helmholtz resonance was likely responsible for their production. No signals occurred without this neck stretch or the less forceful 'head throw' behaviour in which the chin is lowered and then raised so the nose is pointed straight up. (These two behaviours are also present in okapis.) When giraffe were separated, neck and head movements indicating infrasound production were reported more often from adult males and young females than other individuals. Were females alerting males that they were in oestrus?

The long wavelengths of 20–40 Hz favoured by giraffe mean that signals are reflected only by very large objects, so they are ideal for long distance communication. Researchers have long thought that giraffe are not very social because their herd mates change frequently day by day, but they may instead be very social indeed, keeping in close touch by low vocal sounds in a way that completely eludes us. Perhaps individuals keep track of their buddies assiduously, even when they are far apart, like teenagers with cell phones. It would be fascinating to do simple replay experiments in the field to see how individual giraffe react when an out-of-sight fellow gives a low sound: what exactly are giraffe saying to each other?

Senses

Giraffe exist in their environment in the same way we do: through their senses. Almost no experimental work has been done on these in this species, so this summary will be brief and general. In part because of their great height, giraffe have excellent eyesight on which they largely rely to warn them of predators or alert them to other giraffe. Dieter Backhaus (1959) did carry out a zoo experiment to find out if they had colour vision. This involved offering a male food in up to four containers, each marked by a different colour including containers in shades of grey with an intensity similar to that of the colours. He believed that the giraffe were able to distinguish between red, orange, yellow, yellow–green and violet. Yet sight is not essential, at least for a juvenile: George Schaller (pers. comm.) encountered a mother and her 3- or 4-month calf who was blind, yet in good condition.

Hearing is also highly developed in giraffe. Any noise that I noticed while in the veld (far-off laughter, a car in the distance, a sable antelope's snort) was heard also by nearby giraffe who turned (like me) to look in the direction it was coming from with their ears facing forward. They also, of course, hear sounds too low for human ears.

If they have an itch, giraffe may rub their bodies in bushes to relieve it, so they obviously have a sense of touch, but we know almost nothing about their other two senses, taste and smell. Presumably these are keen, because they must use them when feeding to choose what plants, and what plant parts, to consume and which to leave alone. They are often seen to touch another giraffe, especially a youngster, with their nose, as if smelling it. Some biologists believe that giraffe have an excellent sense of smell, but others disagree (Dagg and Foster, 1982).

Giraffe are very curious, perhaps a sixth sense? Once when I was in the veld waiting in my small car in the noon heat to carry out a census every 5 min of what the giraffe in the far distance were doing, I decided to pass the time practising ballet exercises beside the vehicle. Even before the next 5 min were up, a female giraffe was walking slowly toward me, staring intently. What on earth was I doing? I had to retreat into the stifling hot car so I could continue the survey.

Flight distances of giraffe, the moment they decide to flee because a possible danger has come too close, are highly variable. When they were hunted almost to extinction 100 years ago, they fled if a man came within 450 m of them (Madeira, 1909). Today, they don't bother to look up if a jumbo jet sweeps low overhead on approach to the Nairobi International Airport. In the field, their flight distance varies with such things as the threat, past experience with people or lions, the number of giraffe nearby, their mood, the time of day, the terrain, the vegetation and the weather. Giraffe will tolerate a vehicle near them far more than a person on foot. Indeed, those that live in enclosed areas may become so used to cars that they will walk right up to them and sometimes give them a good kick (Dagg and Foster, 1982).

Not swimming

Two scientists have recently calculated mathematically that although giraffe have never been known to swim, theoretically they could manage (although to my mind this would take a miracle) (Henderson and Naish, 2010). In deep water an adult would have its neck stretched out subhorizontally with its chest pulled down by its large front legs. In this awkward position it couldn't move its neck back and forth in tandem with its leg movements as it does during walking. However, its body is buoyant in water deeper than 2.8 m. If it knew enough not to struggle and thrash about but to relax, presumably it might survive for a while with its head above water.

Two giraffe have drowned under unusual circumstances. One who was being unloaded at New York docks broke free from its crate, ran down the deck and plunged into the water. It sank from view immediately without apparently attempting to save itself (Crandall, 1964). Another animal observed lying in a muddy river in Kruger National Park might have regained its feet, but it was unwilling to put its head under water to effect this and eventually drowned (C. S. Churcher, pers. comm.).

One giraffe does, however, like to take a dip in the swimming pool of the Kilimanjaro Golf and Wildlife Estate in Tanzania (Enoch, 2012). Monduli, now a young adult, was rescued as a baby by the anti-poaching unit of the Tanzanian Wildlife Department. He wades (walks?) the length of the pool without having to worry about keeping his head above water, and actually leaps out of the pool when his dip is over. (To the consternation of other players, he also likes to be involved in games of football and polo.)

6 External features

Size

Why is it great to be big and tall? Unlike other browsers, giraffe eat leaves growing near the ground, leaves 5.5 m (18.2 ft) above the ground and leaves everywhere in between; a large male measures about 4.5 m high in a normal standing posture, but it can add a metre to its height if it spies a tasty morsel by tilting its head back and extending its prehensile black tongue. Only elephants and arboreal species can forage so high. Because of their long legs giraffe can walk long distances in search of food or water. They can spot approaching danger sooner than can smaller animals, but have few predators anyway because of their size. This large size allows for efficient use of food, both in its storage and in its fermentation. It means they live a long time with the concomitant beneficial accumulation of experience and of antibodies against diseases. It enables them to keep cooler in hot weather than can small mammals which have to seek shade on scorching days: its elongate shape gives it large expanses of body surface from which internal heat can be dissipated. Giraffe prosper in hot climates without resorting to lolling in water or wet mud as do other heavy megaherbivores – hippos, elephants and rhinoceros. Over 5 months at the Taronga Zoo in Sydney, Australia, I made a note of which animals sought shaded areas on hot days (Dagg, 1970a). The giraffe was one of the few species which seemed impervious to the heat; only on the hottest day (54°C in the sun and 38°C in the shade) did most of the herd of 18 choose the shade. Even so, six animals remained in the sun.

Hide

The hide or skin of the giraffe has features related to their dermal armour, blood pressure and thermoregulation. To study these adaptations, Farzana Sathar, Ludo Badlangana and Paul Manger (2010) examined in detail the complete skin of one young male and confirmed their findings on the skins of two other males, one young and one adult. They found that the skin was often thick, measuring up to 20 mm, but that its thickness varied substantially across the body surface. The outer epidermis was very thin, overlying a thicker superficial dermis and a very thick deep dermis. Sweat glands (which are inactive) and larger blood vessels demarcated the two dermal layers, both composed mainly of collagenous fibres. There were also significant densities of elastic fibres in areas of the superficial dermis.

The skin was thinnest on the head, slightly thicker on the legs, and thickest in the upper half of the trunk and rump and on the anterior and lateral surfaces of the neck. The authors suggest that the thickened skin on the neck may have a protective function when the males battle by hitting each other with their heads, but that their bodies and rumps are not involved in combat. However, this is incorrect. Fights very often do result in blows to the rump or back of an opponent. A thick body skin may also protect against lions that leap onto a giraffe's back in an attempt to bring it down (see Chapter 4) and against the thorns of trees among which it browses.

Blood pressure challenges are a feature of giraffe because of their height (see Chapter 8). The heart has to push arterial blood up to the brain far above it, and the legs must return venous blood to the heart, again, far above. To facilitate these anti-gravity needs, the inflexible collagen fibres in the dermal layers of neck and legs form in effect tight stockings similar to tight anti-gravity suits designed to maintain blood flow that are worn by fighter pilots at high-acceleration gravitational forces. The giraffe's movement is not constrained by its skin, however. In body areas where there is movement over joints, the two dermal layers have elastic fibres that facilitate this. For example, the superficial dermis of the rump contains a high density of elastic fibres apparently related to hip movements, while the deep dermis of the legs has some elastic fibres for movement purposes, but still much collagen for the rigidity needed to maintain blood flow.

Farzana Sathar, Ludo Badlangana and Paul Manger (2010) corroborate the findings of Mitchell and Skinner (2004) about thermoregulation. There is a dramatic difference in the surface temperature of the spots (= the darker, thicker patch skin) as opposed to the lighter and thinner surrounding non-patch skin, reflecting anatomical differences because there are more blood vessels in the patch skin.

Recently a new skin disease to be called Giraffe Skin Disease (GSD) has been described from Ruaha National Park, the largest in Tanzania, affecting 80% of giraffe in all areas of the Park and manifesting as chronic and severe scabs, wrinkled skin, encrustations and dry or oozing blood on the front legs. It is more prevalent in the wet than in the dry season, causes limping in some individuals and adults are significantly more affected than young animals. This debilitating ailment makes giraffe much more vulnerable to lion attack, drought or fire, so officials are anxious to study it in depth (Anon., 2012e). How is it spread? Perhaps one giraffe can catch it from another, or from oxpeckers which may also spread Giraffes Ear Disease (GED), but where did it come from originally? From biting insects? Spores in the grass? No one knows. Research is badly needed to ensure the future of the Ruaha giraffe (Epaphras *et al.*, 2012).

General colouring

All giraffe have spots or patches on a paler background which retain their shape throughout life and which may reflect their race (such as reticulated for the reticulated race and jagged for Masai giraffe), or may not, as giraffe from different races can look almost the same.

Information from zoos about the colouring of giraffe is confusing. At the Taronga Zoo in Sydney, calves of mixed race usually had light brown spots, but on one male of

Figure 6.1   Prime Rothschild's male on the left has dark areas in the centre of each spot. The old male behind him has almost black spots, three large horns and facial wrinkles. Photo by Mika Bona.

6 months they were almost black; these calves also had a black stripe running down the front of each foreleg which lightened as they grew (Dagg, 1968). Some of their spots had pale or white central areas that darkened as the calves matured. Indeed, most spots darken somewhat with age, with extra pigment laid down in a spot's centre gradually spreading outward; in 24-year-old male Jan, the spots were almost black. They may be similarly dark, too, in wild male Rothschild's giraffe. Males of this race may have dark centres in their spots (Figure 6.1) while photos of reticulated giraffe show that some have white central areas in their spots.

There is some evidence that colour may be inherited. It would be of interest to research in depth in captive giraffe the different colouring of spots and background (usually tan or cream): in individuals themselves over their lifetime; in how pigment is laid down in spots; and in the shape and colour of spots and background from parents and their young and from individuals of mixed race. I made a first attempt long ago (Dagg, 1968).

Recently, research on giraffe in the wild indicates that for males, the older and more dominant they are, the darker their coats. Rachel Brand (2011a) has found for *angolensis* males in Namibia, that:

(a) older, taller males have darker coats than other giraffe;
(b) only darker males were involved in intense fights, but never in sparring;
(c) darker males usually won such fights;
(d) when such a dark male came along, other giraffe stepped aside;
(e) darker males were more likely than others to elicit urination from a female; and
(f) most courtships of females and all observed matings included dark males.

Brand concludes that a dark pelage in males is a reliable sign of high status, social maturity, competitive ability and probably reproductive success. (See aging of large males by their black spots in Chapter 9.)

Rarely, giraffe are born with unusual colouring. One brown giraffe with no spots was born in a Tokyo zoo of parents of normal colouration, while others have spots that are faint. Dagg and Foster (1982) give 16 references from the literature of white giraffe seen in the wild. One white Masai animal was probably an albino with concomitant poor eyesight because it kept near the centre of a group for safety, but others were not because they had dark eyes. Because of these variations in wild animals, it seems that coat colour and pattern is not under strong selection pressure by predators as may be the case in smaller species. (A white giraffe with darker spots is pictured on http://www.crystalinks.com/giraffe.html.)

Giraffe have spots so these surely serve to some extent for camouflage, as early writers have hypothesized. Perhaps giraffe immigrating into Africa from the northeast were monochrome, as some prehistoric Saharan drawings depict? Perhaps polygonal spots evolved to emulate open-angle polygonal branch arrangements of trees upon which giraffe feed such as *Acacia giraffae*? Perhaps the reticulated giraffe in Kenya have sharp-edged spots to match sharp, bright equatorial shadows, while southern giraffe have less well-defined spots to emulate the more diffuse shadows there? Perhaps the rich colouring of reticulated giraffe and of Rothschild's giraffe harmonize with the bush and scrub among which they live, while the shading of the relatively light-coloured sahel and Angolan giraffe is paler to match the sandy shades of semi-desert regions? There is as yet no way to know.

Z. Yong (1994) has found that the coat pattern of the giraffe is consistent with theoretical treatment of topological constraints upon inhomogeneous surface networks of solids.

Spot patterns of different races

For part of his PhD thesis, Russell Seymour (2001) decided to determine once and for all if the spotting of the various races of giraffe was distinctive enough to act as a marker for each. All the original descriptions of giraffe spotting of a new subspecies/race indicated that coat pattern was a diagnostic characteristic, but all too often the coat of giraffe of one race looks very like that of another.

On museum visits to a number of countries, Seymour laid out 72 complete or partial skins of giraffe of 8 known races so that he could photograph each from a consistent height to measure and describe later for comparative purposes. Next he listed 40 characteristics that might apply to each skin (such as spots between the eye and ear, width of rump lines, and size distribution of neck spots etc.). By coding the peculiarities of each pelage, he came up with a number of interesting findings that could be applied to the eight races he was comparing. (He did not have information on the skin of the rare Nubian giraffe, and two races were represented by a single skin, so further research should centre on such groups.)

Seymour found that:

- the spotting of *reticulata* and of *tippelskirchi* was highly diagnostic;
- the pelage of *peralta* is fairly distinctive with its broad inter-spot lines, especially on the neck;
- the spotting of two groups from southern Africa (*giraffa* and *wardi*) appear to be the same (and are now classified as such, with *wardi* deleted);
- there is a trend for spots to occur on the head and face of northern races, but not on those of southern races, so this could be a valid diagnostic character to be used in combination with other characters;
- the shape of spots can, of course, be highly diagnostic, but the spaces between spots have little taxonomic value (except for the reticulated race); and
- spotting below the hocks has been used in the past as a descriptor for races, but it is individually variable within geographic specimen sets, so useless as a taxonomic character.

Seymour reports (pers. comm., July 2012) that because of his extensive research on coat patterns, he was able to tell a giraffe's race from its pelage most of the time. When a group of sceptical friends doubted this and for fun put together a slide show of giraffe photos he had never seen before, he correctly identified the race of over 70% of these animals (but of course was wrong about 25% of the time).

Anthony Hall-Martin reported that spot patterns were already present on a 100-day-old giraffe foetus, although other zoologists believed that they appear only much later (Murray, 1981). J. D. Murray noted that 'the vasculature of the skin is highly organised in relation to the pelage pattern with superficial arterial vessels surrounding each patch' (p. 171). However, these arterial outlines disappeared quickly after the death of the foetus which is presumably why other anatomists did not see them.

Hair

Several studies have concentrated on the growth of hair in giraffe embryos. Two embryos of 3 months (28 cm body length) had no visible hair, while another of over 4 months (60 cm body length) had hairs on its lips, around its face and on its tail tip (Beddard, 1906; Broman, 1938b). One embryo of 8 months' gestation had very fine hair on most of its body but was of uniform colour, with no spots at all. The hair of full-term newborn giraffe is soft, slightly woolly, short and of course coloured to make up spots on a lighter background.

A hundred years ago, unlikely as it may seem, academic attention focussed on the whorls, feathering and crests of hair on the bodies of giraffe (Kidd, 1900, 1903; Rothschild and Neuville, 1911). This was because W. Kidd believed that their arrangement supported the belief of Lamarck – that evolution proceeded not by natural selection as Darwin had explained, but by the inheritance of acquired characteristics. He argued that the direction of the hair was correlated either with attitudes of rest, activity or locomotion in various regions of the body. For example, joint hairs are affected every

Figure 6.2   Male Rothschild's giraffe at Soysambu, Kenya. Photo by Zoe Muller.

Figure 6.3   Male Rothschild's giraffe at Soysambu, Kenya. Note the large difference between the markings of this giraffe and that in Figure 6.2. Photo by Zoe Muller.

time a giraffe gets up or lies down so that tufts and whorls soon appear near the joints involved in these exertions. The exciting finding was that these hair arrangements were present even in embryo giraffe which had never made movements similar to those of living animals. Kidd argued that because Darwin's natural selection would never occur for such a minor feature as the direction in which hairs lie in embryo giraffe unless there had been divine intervention, the hairs of the embryo emulated those of its parents; thus acquired characteristics had passed into a second generation as Lamarck had predicted. Later zoologists have not followed up this line of thought.

Early zoologists discovered that the slope patterns of hair of okapis as well as of giraffe were more like those of the Cervidae than any of the other ungulates, indicating perhaps a close relationship between these two families (Lankester, 1902, 1907, 1910). The hairs of the giraffe's neck mane are directed backward while those on the middle of the back are directed forward. The crest where these two currents meet varies in individuals but is much farther back in the giraffe than it is in other ruminants, undoubtedly because of the greater encroachment of the neck on its trunk. The hair on the head of giraffe is differentiated into parallel tracts of dense hair that stand upright and intermediate tracts of smooth flat-lying hair; there is a suggestion that the patterns of hair on the head may vary with different races, but this needs to be researched.

The microscopic structure of the hair of giraffe – from the flank, wrist, mane, tail and eyelash – have been described by T. Lochte (1952) and Broman (1938a), while hair whorl patterns are reported for 26 reticulated giraffe by L. J. Dobroruka (1976).

It has been assumed that giraffe do not thrive in cold weather, given in the past that their range did not extend as far south as the Indian Ocean. However, they do live now in game farms and reserves in southern areas of South Africa. Zoo-man Carl Hagenbeck (1960) confounded experts who felt that captive giraffe, like all tropical animals, should be kept in warm quarters. Instead, he housed them in unheated stables in Germany where the temperature was often only a few degrees above freezing. The animals flourished, emerging in the spring with a thick crop of hair two and a half times the normal length (Street, 1956). However, it now seems apparent that cold weather actually causes in part the condition of Serous Fat Atrophy that suddenly kills many zoo giraffe without warning (see Chapter 10).

Tail hair

The tail is useful to the giraffe as a fly switch, keeping away biting and other insects from its body. The long thick black hairs at the end twist over its back as a giraffe runs, perhaps so they do not get mixed up with the moving legs, but otherwise they hang down loosely. We know about the structure of giraffe tail hairs because they resemble those of African and Asian elephants (Yates et al., 2010). Both of these pachyderms are listed as endangered, threatened or in need of protection in some or all areas by CITES (Convention on International Trade in Endangered Species of Wild Fauna and Flora) and by ESA (US Endangered Species Act). (By contrast, giraffe, even the peralta race in Niger which includes few members, are not protected under CITES or ESA.) To enforce the mandates of CITES and ESA, and because elephant tail hairs are similar to those of giraffe, the hairs of both species have been carefully analysed so that wildlife conservation officers will know which bracelets, rings and earrings are made of elephant tail hairs and therefore illegal.

Even under the microscope, the coarse black tail hairs of these two species can be confused (Yates *et al.*, 2010). Microscopic cross-sections of these hairs show that their structure more closely resembles that of the rhinoceros horn or horse hoof than that of their other body hairs. Their make-up is of tiny keratinous cylindrical tubules tightly packed into a tail hair with pigment granules in the matrix of each whose positions and densities differ according to species. In the giraffe tail hairs the tubules are circular or oval with a distinct radial arrangement. Light microscopy indicates that cross-sectional shape, pigment placement and pigment density are all useful morphological features for distinguishing these species. There are also chemical differences between elephant and giraffe tail hairs (Espinoza *et al.*, 2008).

Ears

The ears of the giraffe are highly mobile, facing down at rest, or flared out and forward when an individual is concentrating on something of interest in front of it. In the Mikumi National Park, Tanzania, there has been a spate of ear infections called Giraffe Ear Disease (Mlengeya *et al.*, 2002; Kagaruki *et al.*, 2003; Epaphras *et al.*, 2012). Little is yet known about it, but it could be spread by oxpeckers flying from the head of one giraffe to another, or by flies seen on the skin lesions of affected animals.

Horns

Both male and female giraffe have what are referred to as 'horns' (really ossicones) on the top of their head, but they are not the same as horns present in members of the Bovidae family. Rather, they are undifferentiated bony growths on the skull covered with hair, those of males from 10 to 25 cm in length and those in females smaller and more slender. Ancestral giraffids had larger and more ornate horns than giraffe (see Figure 1.1), but these would have been too heavy and unwieldy in combat for an animal with a long thin neck. The main horns lean backwards rather than standing erect, those in the male sometimes growing so large that they meet at their base, thus obliterating the median suture (Singer and Boné, 1960). The two horns are usually asymmetrical in any one animal in that they may lean at different angles and may differ in length by as much as 35% (Rothschild and Neuville, 1911).

When a calf is born it has horn cartilages lying flat under the skin of the head (to facilitate the birth process) which stand up within a few days under the thick black hair tufts which mark their place, giving him or her a startled appearance – giraffe are the only animals to be born with horns (Figure 6.4). Later they become ossified and fused to the skull over the parietal and frontal bones. During a giraffe's lifetime bony tissue is laid

Figure 6.4   Neonate calf's wonky cartilage horns soon ossify and fuse to the skull. Photo by Zoe Muller.

down in front of the two robust main horns, forming what has been called a third horn that is especially prominent in males. Other bony growths may occur on other parts of the skull as discussed below. These head growths in male giraffe help cushion the brain during head-hitting contests (Dagg, 1965). Sparring and fighting among males is so common that the black hair sticking up around the horn in young animals is completely worn off in adult males. Thus one can tell the sex of an adult immersed in vegetation by its horns. (In the Taronga Zoo, however, there was an unusual female, 11-year-old Cheeky, who never bore young and who often head-battled with each of the three adult males, always losing to them. The hair on her horns was also worn away; Dagg, 1968.)

Some giraffe possess what have been called 'mizzen' horns at the lower back of the head that serve for the attachment of posterior neck muscles and the *ligamentum nuchae*. As they were prominent in the two older males at the Taronga Zoo and quite large in the youngest 5-year-old male and in Cheeky, their size is probably related to the frequent sparring of the four animals. The exertions of the neck muscles account for the enlargement of the bony growths at the points of insertion. (Cheeky did much less hitting than the males, but far more ducking which also exercised her neck muscles.)

Some authors have suggested that the horns of the giraffe are covered with vascularized skin analogous to the summer antlers of deer as a response to temperature; heat can be released from the body to the air, thus cooling the animal (Spinage, 1970). However, the skin over giraffe horns is thick, like that on the rest of its body, and is not vascular like the velvet skin of growing antlers.

Few spontaneous tumours have been reported in giraffe, but one has involved the horns of a male, the first to be described (Woc-Colburn *et al.*, 2010). A three-year-old Rothschild's giraffe had such an embryonal rhabdomyosarcoma beside and behind the left parietal horn which invaded and destroyed the underlying parietal bone and adjacent horn base and sinuses. The mass was removed and chemotherapy administered, but the animal soon died.

Thinking about evolution, should the unusual horns of giraffe be considered as a primitive type from which more complex horns have developed in other families (Anthony, 1928–29; Colbert, 1935a)? If so, probably the free horn versions of the calf united earlier and earlier with the skull until they appeared to be its original growths. Or are they a degenerate type which were more fully developed in ancestral giraffids (Thomas, 1901)? It could be the former because *Helladotherium* is hornless, but it could be the latter, as *Sivatherium* and *Bramatherium* have more complex horns. In all the fossil giraffids the horns of females are either lacking or less well developed than in the male, indicating that they had sexual significance and that their growth was perhaps proportional to their use. As the modern giraffe evolved to be taller, the neck would increasingly have trouble supporting a heavy mass of horns should these have been present in its ancestors.

Horn research?

A unique feature of giraffe skulls is the tendency in adults to lay down new bony growth, especially in males. Research on the horns and facial bumps of captive giraffe of known age and race and what stimulates their growth would be of interest. Might different races or subspecies of giraffe have different configurations of horns and exostoses? If so, this would strengthen the reality that because some races have actually been isolated for a million years or more, they deserve full species status.

Here are some conjectures relating to race and horns in males.

- A Cape giraffe's skull had large bony growths over each eye, under the left zygomatic arch, and on the right jugal bone as well as small supernumerary bumps scattered over the skull's surface. These growths increased the volume of the horns (Rothschild and Neuville, 1911).
- A Kenyan giraffe had a large orbital 'horn' above the right eye with the capped appearance of a distinct epiphysis. Similar growths may also occur above the left orbit (Lydekker, 1904).
- Such 'right-headedness' has been reported for males of the *rothschildi* and *tippel-skirchi* races (Granvik, 1925).
- Rothschild's giraffe were said to have five horns: the main pair, the third median growth in front of them and two mizzen, occipital or posterior horns at the back of their skull (Thomas, 1901).
- Northern giraffe were reported to have especially well-developed epiphysial horns and orbital 'horns' while southern races had rough scattered secondary growths particularly on the frontal and nasal bones and on the occipito-parietal crest (Rothschild and Neuville, 1911). East African giraffe had an intermediate stage between the more extreme north and south types.
- Skulls of Thornicroft's giraffe were more robust than those of most other subspecies although the parietal 'third' horn was smaller (Seymour, 2001).

This secondary bone growth is of two types: bone laid down over parts of the skull or growths with their own epiphysial centre which later fuse with the skull. Both seem to be a secondary sexual phenomenon which originated either spontaneously or as a result of mechanical irritation from fighting. Yet sometimes females who don't fight also have secondary growths (Urbain *et al.*, 1944; Krumbiegel, 1965). While adult male skulls are rough and heavy, those of female and young giraffe are in general smooth and thin; indeed, the female skulls are more fragile than those of most ungulates of the same size.

Spoor

The hoof print of a large male measures up to 31×23 cm. Those of calves and young bulls are narrower, while those of old males where the hoof halves have spread are perhaps even broader (Dagg and Foster, 1982). The front hooves tend to be wider than the hind hooves, given that the forequarters of giraffe are larger than the hindquarters and bear more weight.

Giraffe droppings resemble large acorns up to 4 cm long, with one end of the pellet flattened. A dropping from a wild giraffe when fresh is olive green in colour, but it shrinks when it dries, turning yellow brown with a dark brown brittle skin that cracks into a network of tiny squares.

Scent

Giraffe have apparently evolved a way to deter harmful microorganisms on their bodies. Some individuals emit a body odour that people in the field have detected up to 250 m downwind (although neither Foster nor I ever noticed it). Curious about this, William Wood and Paul Weldon (2002) decided to analyse the chemicals involved. They put a male and a female captive-born reticulated giraffe in a squeeze stall so they could collect hair and epidermal samples from their neck area; the skin specimens involved scraping up tissue using the edge of a glass slide. The material was placed in glass vials, stored frozen and then extracted with dichloromethane for analysis. Using gas chromatography–mass spectrometry methods, the extracts were found to contain 11 significant volatile compounds which probably come from the widely distributed sebaceous and apocrine glands of the skin, although this was not verified.

The two alkaloids present in the skin samples, indole and skatole (3-methylindole), were largely responsible for the giraffe's body odour. Both are used in perfumery in very dilute solutions, recalling scents of jasmine and orange blossoms (although in high concentrations they smell like faeces). These and others of the compounds may function as antimicrobial agents. Indole, for example, is used to treat athlete's foot fungus and other human skin infections. A related skin fungus, *Trichophyton rubrum*, is inhibited by octanal, nonanal and tetradecanoic and hexadecanoic acids; such acids inhibit the zoophilic fungus *Microsporum canis*. In addition, the four compounds and others inhibit two ubiquitous species of skin bacteria: *Staphylococcus aureus* (causing human boil and staph infections) and *Propionibacterium acnes* (causing acne in people). The researchers note that the antimicrobial activity of most of the 11 compounds is moderate, but acting together it would be more potent.

Other possibilities are more tentative. Both indole and skatole may repel mosquitoes (*Aedes* spp.), and a tick present in areas inhabited by giraffe, *Rhipicephalus appendiculatus*, is repelled by the compound *p*-cresol which is present in both skin and hair of giraffe.

7　Anatomy

Giraffe are big. All races have similar measurements, with adult males in the wild up to 5.5 m tall and adult females about 4.3 m (Bush, 2003). Taller specimens have been reported by hunters who have shot them, but their calculations may reflect a bit of wishful thinking. Giraffe in zoos are usually somewhat shorter, presumably because of their unnatural living conditions. Adult males in the wild weigh from 850 to 1950 kg, while females weigh from 700 to 1200 kg (Bush, 2003); an average weight for males is generally considered to be 1200 kg and for females 800 kg.

When creatures as strange-looking as giraffe were first brought to Europe, men with a scientific bent began to describe them and their internal organs for academic journals. The giraffe's long neck and legs especially have interested anatomists for over 170 years, with Richard Owen publishing anatomical data on the Nubian giraffe as early as 1841.

Many organs and systems of the giraffe have been described in detail in the literature, as this chapter indicates, but others have not. There may therefore be an imbalance in what is discussed here, but this cannot be remedied at present.

Colour photographs of a giraffe dissection are given in http://whatsinjohnsfreezer. com/2012/12/02/thank-you-giraffe-anatomy/. They feature jugular vein, nuchal ligament, cross-section of trachea, heart, rib cage, rumen, molar teeth, inner cheek spines, brain, tapetum lucidum of an eye and a misshapen hoof. These organs/areas are marked with an asterisk* in the text.

Information on the amazing circulation and reproduction systems of giraffe, which have attracted the attention of scores of curious scientists, is covered in Chapters 8 and 9.

Skeletal system

Bones

O. van Schalkwyk, John Skinner and Graham Mitchell (2004) from South Africa and the University of Wyoming joined forces to study the bones of giraffe compared to those of buffalo. Both are African artiodactyls of a similar large mass, but they differ markedly in that the giraffe has a long neck and long legs and browses, while the far more compact buffalo grazes. The giraffe's elongated cervical bones account for the neck length, and its long metapodials (lower leg bones) for the leg lengths which must withstand high loading forces because of their slim size.

To obtain skeletons to study, they collected seven giraffe and nine buffalo carcasses from Kruger National Park: adult male animals which had been killed by predators or culled by hunters. After the neck and leg bones had been separated from the carcasses, they were cleaned of soft tissue, measured, weighed, sometimes cut and had their density calculated by immersing them in water.

The bone densities of the mid-shafts of the femurs and the metacarpal bones had similar values in both species. Significantly less dense in both species were the carpal bones, ribs and vertebrae, with the femur heads least dense along with the vertebrae. The density of the cervical vertebrae were similar in both species, but the mass of these bones in the giraffe was 4–5 times as great as the mass of their lumbar vertebrae, while for buffalo the difference between these two types of vertebrae was only 1.5 times. The densities of the bones for both species were similar, except for the slightly more dense lumbar vertebrae of buffalo. (In another study, the thoracic limb skeleton of a giraffe is compared to that of cows: Damian *et al.*, 2012.)

It had earlier been assumed for giraffe that the relatively low density of cervical vertebrae (to keep the neck light and mobile) and the relatively high density of their limb bones (to support the heavy body) were unique adaptations of giraffe (Mitchell and Skinner, 2003). However, this is not true. What then makes the giraffe limb bones strong? One thing is their straightness, which greatly exceeds the straightness of leg bones in 10 other artiodactyls (Biewener, 1983). The giraffe leg bones are also wider than those of the buffalo (although not as wide as would have been expected from their much greater length) with a narrower marrow cavity (van Schalkwyk *et al.*, 2004).

The low density of the giraffe neck bones (along with the femur heads) makes them lighter and their possible remodelling more feasible with growth. The lower cervical vertebrae seemed to have more mass than the higher ones (only C3, C4 and C5 were measured), as might be expected, so that the neck is largely supported at the base with the top part more manoeuvrable for browsing or hitting (in the male).

These scientists carried on their work with giraffe and buffalo, this time studying the calcium and phosphorus in their bodies which were thought to be finely balanced in these species (Mitchell *et al.*, 2005). The two chemicals, calcium and phosphorus, were of especial interest because:

(a) rickets and kidney stones have been linked in giraffe to a Ca/P imbalance (Langman, 1978);

(b) osteophagy in giraffe is a possible sign of mineral deficiency, especially of P;

(c) the giraffe skeleton elongates faster as it grows than it does in any other mammal; and

(d) their skeleton constitutes a greater proportion of their body mass than it does in, say, cattle. In cattle, the skeletal mass is about 15–20% of the carcass mass whereas it is somewhat higher in giraffe. Daily Ca absorption is estimated over a 5-year growth period to be about 29 g/day in giraffe. The giraffe skeletons contained more Ca and P than did the buffalo skeletons, presumably because the Ca content of browse consumed by giraffe averages about 40 g of Ca per 20 kg per dry matter, while grass gives buffalo only 8 g per 20 kg. Giraffe have to accumulate 1.5–2 times as much Ca during growth to maturity as do buffalo.

The density of the bones was correlated significantly (but not entirely) with their Ca and P content. Bones within a species vary in their mineral content – in giraffe there is more Ca and P in the strong leg bones than in cervical bones, for example. Giraffe have a relatively rapid growth rate during which they need more Ca and P than do buffaloes. Ca is available from browse, but P is more difficult to obtain.

Giraffe obviously need good sources of calcium and especially phosphorus to sustain growth and maintain their large skeletons. Might osteophagia (the consumption of old bones) contribute to giraffe nutrition? I. P. Bredin along with Skinner and Mitchell from the above studies (2008) took on this question. Giraffe often reach to the ground to pick up an old bone to chew. Are they bored? Is the bone tasty? Is this just a habit? Or does it supply phosphorus the animal may be lacking? Earlier research had shown that osteophagia in cattle could be eliminated by supplementary feedings of P in the form of bone meal (Theiler *et al.*, 1924).

The researchers posed the following question: can Ca and P ingested as bone be released in the rumen in a form that can be absorbed into the body in the lower intestine? Samples from cancellous (cervical vertebra) and dense bone (metacarpal shaft) from the studies just described were cut into small cubes, defatted, oven-dried, placed in small nylon bags and hung either in distilled water, artificial saliva or rumen fluid in fistulas of five mature sheep. They were left in place for a month. Water had no effect on the bones. The dense bone samples were softened by exposure to saliva and rumen juice, but did not lose Ca or P. The cancellous bone also softened and their mass and volume decreased, but also lost little Ca or P. The researchers concluded that insignificant digestion had occurred in the rumen and therefore osteophagia does not lead to better nutrition.

Michigan State University researchers have produced a baseline assessment of bone density and mineral elements in giraffe and other mammals, describing variations in bone ash residue and bone phosphorus levels (Middleton *et al.*, 2009).

Skull

Anatomical differences in the shapes of skulls are an important part of taxonomic work. However, although many measurements have been taken of giraffe skulls, there has been little attempt to correlate differences with racial characteristics (Rothschild and Neuville, 1911; Dagg and Foster, 1982). Such research, if a correlation were found, would be helpful in backing up the contention that many giraffe races are actually full species.

The adult skulls of males are much sturdier and heavier than those of females because they act as a weapon with which to lambast other males during sparring or dominance combats. Three such male skulls with mandibles weighed on average 9.9 kg (range 8.9–10.9), while the average weight of seven female skulls with mandibles was 3.5 kg (range 2.4–4.7) (Dagg, 1965).

The part of the giraffe skull in front of the eye sockets is especially long, apparently because of its diet of arboreal vegetation (Godina, 1970). The eyes themselves are at the side of the head, giving the giraffe a wide field of view, but not binocular vision as we ourselves have. The facial bones are flat, with the nasal opening long and narrow. As a youngster reaches adulthood, the area around its eyes and upper nose widens (Singer and

Figure 7.1   Skull of male Rothschild's giraffe. Note five 'horns'. Photo by Zoe Muller.

Boné, 1960). The nasal region is raised, with overlaid thickened layers of bone and often in males a median horn formed from a separate centre of ossification (see Chapter 6). During maturation the cerebrum becomes relatively larger in proportion to the brain case (Cobbold, 1860) and shifts from a position above the level of the nasal passage to one level with it. The auditory bullae are relatively larger in young giraffe than in adults (Colbert, 1938).

The occipito-parietal crest at the back of the skull is well developed even in young animals; below it is a deep furrow where the strong neck muscles are attached (Rothschild and Neuville, 1911). The prominent occipital condyles are well rounded, allowing the skull free movement; for a show of dominance when fighting, or when feeding at a high tree, a male giraffe can tilt his head until it is pointing straight up.

As an animal grows, the skull bones gradually fuse together. This happens first for the interparietal suture on the top of the head, about the time a calf is born (Singer and Boné, 1960). Soon after the parieto-occipital suture at the back of the skull fuses, followed by the occipital suture by the time an animal is four, and the fronto-sagittal suture soon after. As a giraffe becomes fully grown at five or six years, the fronto-parietal suture across the top of the head closes, followed much later by the temporo-occipital and palatal sutures.

Giraffe skulls have various sinuses, in part presumably to lighten their heads somewhat that are perched on such long necks. These sinuses are bounded and transected by fibrous connective tissue walls as well as by bony partitions, so their structure should be studied

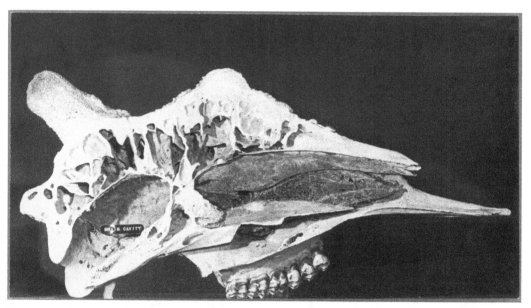

Figure 7.2   Internal section of skull of male Masai giraffe. Note heavy dorsal bone structure.
Photo by Bristol Foster.

on wet rather than dry specimens where the tissue wall will have been destroyed (Anthony and Coupin, 1925). Skulls include small anterior nasal sinuses, larger paired posterior nasal sinuses, and well-developed sphenoid sinus pairs and superior maxillary sinuses. However, the most impressive is the vast frontal sinus.

Frontal sinus

Ludo Badlangana, Justin Adams and Paul Manger (2011) tried to make sense of why giraffe have an exceptionally large frontal sinus in their skulls. They did this by studying the bony cranial mass and frontal sinus volume of the cranium (the part of the skull that covers and protects the brain) of six giraffe of varying ages, along with single crania of okapi, dromedary, gnu, warthog, domestic cow, tapir, bush pig and domestic pig for comparative purposes. Each of the crania was CT scanned for analysis.

The frontal sinus of both male and female giraffe was relatively larger than that of any of the other species (the next largest was that of the okapi), extending as the young giraffe grew to the caudal end of the cranium, into the occipital bone, and completely overlying the endocranial cavity. The sinus became enlarged because the supraoccipital, parietal and frontal bones grew at a much faster rate than did the rest of the cranium.

Presumably the frontal sinuses in the non-giraffe species are adapted for various purposes such as increasing the vault size for muscles of mastication, for horn support or for the development of secondary sexual characteristics. Why were they so large in giraffe though? The researchers considered various possibilities.

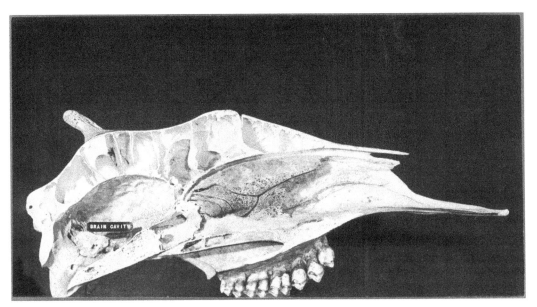

Figure 7.3   Internal section of skull of female Masai giraffe. Note light bone structure and elaborate folding of cheek teeth. Photo by Bristol Foster.

(a) Perhaps the enlarged cavity helps to regulate the temperature of the brain? However, there is no air flow within the sinus itself, which has no direct opening to the nasal cavities. As well, giraffe already have a vascular *rete mirabile* at the base of the skull which helps control temperature.

(b) Perhaps in the giraffe the large frontal sinus helps accommodate the temporalis muscles of mastication? However, careful examination indicates this would be unlikely. A sagittal crest along the top of the skull might work, as is present in the camel, but the giraffe doesn't have one.

(c) Perhaps giraffe produce an infrasonic vocalization for which the extensive frontal sinus acts as a resonance chamber for the production of low-frequency sounds? This could be so, but there is as yet no proof.

(d) Perhaps the enlarged sinus is somehow connected with the horns (ossicones) above, which in adult males become increasingly underlain with an extensive bony table and prominences which protect the brain as a giraffe pounds an adversary with these weapons. This could explain why the sinuses are larger in males than in females, or it may be that females, being smaller, naturally have smaller sinuses.

(e) Perhaps the sinus is large to help reduce the weight of the head which is balanced on the long thin neck.

The reason for the greatly enlarged frontal sinus of adult giraffe is still unresolved, as is whether that of the male is proportionately larger than that of the female. More research is obviously needed.

Teeth

The giraffe has no upper incisor or upper canine teeth; these have been replaced by a hard palate, so that the dental formula of the giraffe is the same as that of cervids, bovids and pronghorn antelope:

0.0.3.3

-------- (incisors, canine, premolars, molars) with 32 teeth in all.

3.1.3.3

However, the giraffe is unique in having an extra lobe or two in the lower canine teeth (posing in effect as another incisor) which widens the row of front teeth, making it more effective in cropping vegetation. Should this canine tooth be discovered on the site of a fossil dig, it would be known to have come from a giraffe or one of its ancestors.

The incisors and canines of the lower jaw are arranged in an arc (Boué, 1970), midway between the nearly rectilinear shape in bovids and the more elliptical shape in camels. A giraffe's front teeth meet the upper callous pad with their lingual face when it closes its mouth, so its grip is less strong than if the incisors were vertical. Therefore, when feeding, it tends to pull leaves from their stems rather than clip the stems themselves. These front teeth each have a single root which extends backwards horizontally.

The longitudinally plicated cheek teeth* of the giraffe are anchored more strongly by two roots at the front and back in the lower set and by three in the upper set – two lateral and one medial. They are low-crowned and broad, nearly square in cross-section, with a rough enamel coat. (Roberts (1951) gives cheek-teeth measurements of four Cape giraffe.) The long diastema (space between the front and cheek teeth) becomes longer in older giraffe as the molar row decreases in size (Willemse, 1950).

The milk teeth comprise 6 lower incisors, 2 lower canines and 12 milk molars, the latter occupying the sites where the permanent premolars will emerge. All are relatively longer than their replacements will be.

The age of a giraffe can be estimated fairly accurately up to its sixth year by the eruption pattern of its teeth (Hall-Martin, 1976):

● in a female born a month prematurely no teeth had yet erupted (Lang, 1955);
● at about 4 weeks of age, all the deciduous teeth are present and worn except for the canines which are erupting;
● at 1 year all the deciduous teeth are worn and the first molar is erupting (thereby extending the row of deciduous cheek teeth);
● at 2 years the second molar is erupting;
● at 3.5 years the third molar is erupting;
● at 4 years the first deciduous incisor has been replaced by the permanent one;
● at 5 years the second permanent incisor and the three premolars are erupting;
● at 5.5 years all the teeth are permanent with the exception of the canines; and
● the canines erupt at 6 years and are fully functional by 6.5 years.

Giraffe older than 6 can be roughly aged using measurements from the mean lingual crown height or mean lingual occlusal surface width of the first molar using Hall-Martin's calculations, by the number of dark-staining incremental lines in the cementum and by the wear pattern of this tooth. The accuracy of these methods is likely within a year for a giraffe up to 15 years old, and within 2 years in older animals. Hall-Martin tried to correlate the age of giraffe with their mandible measurements, characteristics of thin sections of undecalcified teeth and mass of eye lens, but without success. Recently, however, it has become possible to age large males approximately by the darkness of their hide (see Chapter 9).

The way teeth wear down depends on diet. Grazers eat grass which contains abrasive silica whereas browsers such as giraffe eat leaves with little silica. The silica present in a diet scores teeth (defined as mesowear in palaeobiology) in a distinctive way. Marcus Clauss and colleagues (2007), who examined the teeth of 20 free-ranging giraffe and 41 captive ones, found that, as expected, the former were typical of the teeth of browsers while the latter teeth were abraded like those of grazers. They conclude that the common zoo diet of grass hay and pelletted compound feeds contains higher amounts of silica than do leaves, causing abnormal wear patterns for captive giraffe with deleterious long-term consequences.

Enamel hypoplasia of teeth

Enamel hypoplasia is a developmental tooth defect which provides a permanent record of systemic stress during the early life of an individual. It was first used by anthropologists and palaeontologists to assess the health of people in early cultures (Goodman *et al.*, 1980) – were they stressed at particular times in their lives by events such as famine or warfare?

Tamara Franz-Odendaal (2004) has now used the method to compare stresses in the lives of wild and captive giraffe. During upsetting occasions such as weaning, nutritional stress or illness, the growth of cells laying down enamel matrix in a young animal's teeth (ameloblasts) is disrupted, causing a thinning of the enamel known as linear enamel hypoplasia. Other hypoplasia defects may be localized rather than in a layer. Unlike most tissues, enamel does not change during an animal's lifetime unless it is abraded through wear. The wider the hypoplasia layer, the more long-lasting was the time of stress, while its depth is thought to reflect the severity of the trauma. Teeth develop from the tip to the base, so the age at which hypoplasia occurred can be estimated from its position on the crown. To assess the timing and duration of early developmental stress events, one must measure the height and width of linear enamel hypoplasias on the crown surface of a tooth, and know the timing and sequence of tooth development. This method can only be used to rate stress in giraffe when they are younger than 6.5 years of age, after the last teeth have erupted.

For her research, Franz-Odendaal carefully studied 553 teeth taken from 23 giraffe skulls of various sizes and therefore age, noting each layering and defect (if any) on each tooth, its place on the tooth crown and the tooth's position in each jaw (to establish age, knowing the various teeth erupt in a set sequence). She found that the wild giraffe had been relatively stress-free during their first 6 years, as most did not have linear defects in their teeth. For most captive giraffe, however, both males and females, three stress periods were noted: during weaning at up to 6 months of age (noted from defects in the first and second molars); at 3–4 years when males may be subject to bullying in their enclosure and females may become pregnant (noted in third molar and fourth premolar defects); and, the least common with only two animals, at 4–5 years (noted in the second and third premolars) for no known reason.

There are still unanswered questions in this research – why may hypoplasias occur on one side of the jaw and not the other? How, if at all, are localized hypoplasias related to linear ones? Certainly the research findings will be important in assessing posthumously how much trauma wild giraffe are exposed to when they are captured or relocated when young, and how well giraffe calves in captivity fare when subject to specific treatments or new food regimes.

Franz-Odendaal, A. Chinsamy and J. Lee-Thorp (2004) even found enamel hypoplasia on the teeth of the early giraffid *Sivatherium hendeyi* from South Africa which lived over a million years ago. They hypothesize that poor environmental conditions and nutritional distress were likely responsible for this.

Spine

The long neck of the giraffe is so fundamental that one might assume this feature would be evident in the early foetus. This is not so. When an infant giraffe is born, its neck is significantly shorter in proportion to its body than is the neck of an adult, so during an animal's years of growth to adult size, the neck bones grow proportionately faster than the rest of the body. Perhaps the neonate has a shorter neck because during parturition in giraffe and other ungulates the forelegs appear first followed by the head, lying between or alongside them. If the neonate had a neck proportionately as long as that of an adult, its head would appear first, with risk of elbow flexion and/or foetal death.

For their study of bone growth, Sybrand van Sittert, John Skinner and Graham Mitchell (2010) worked with the cleaned neck bones of 39 giraffe of all ages and 9 foetuses, all killed between 2007 and 2009 in Zimbabwe in the name of population management. Each bone was measured for length, width, height and length of its spinous processes. Gender was not a significant cofactor in any of the dimensions (with one exception mentioned below), so the data for males and females were pooled.

As in almost all mammals (but not manatees or sloths), the giraffe neck contains seven cervical (C1–C7) vertebrae. In most ruminants the first thoracic vertebra, T1, is a transitional vertebra in its articulation patterns between the cervical and thoracic vertebrae, but in giraffe this transition occurs between T1 and T2. C7 has become elongated (like C2–C6) and therefore contributes to the neck length, whereas in other ruminant species only C2–C6 do this. T1 has a length half-way between those of the cervical and thoracic series, but is clearly thoracic because it has articulating ribs (like the other thoracic vertebrae) that attach to the sternum. The cervicals articulate with each other by lateral pairs of anterior and posterior articular facets, but the thoracics typically have median pairs because they are less mobile. The cervical vertebrae of an adult giraffe are about 129% longer than the trunk vertebrae (Slijper, 1946).

The dorsal spines on the cervical vertebrae are fairly short, but they increase sharply in length from C7 to T4, where there is a noticeable muscular hump along the back of the giraffe, then decrease slowly into the lumbar region. Their function, especially in the thoracic and to a lesser extent the lumbar area, is to support the head and neck and anchor the important suspensory ligament (*ligamentum nuchae*) of the neck; this ligament has its main insertion on the skull below the occipital crest. The long dorsal spines on the thoracic vertebrae, especially on T1 and T2, are particularly long in large males, presumably as a biomechanical adaptation because of their much heavier head. Of the 14 thoracic vertebrae, the first 7 have true ribs which attach to the sternum while the last 7 ribs are not so attached. The sternum itself is made up of six bones. The lumber vertebrae number 5, the fused sacral bones 4–5 and the tail 16–20 caudal bones (Flower, 1870).

Sometimes a giraffe suffers a spinal injury. This happened to Khalid, a 2-year-old male reticulated giraffe who arrived at the Cheyenne Mountain Zoo in 2010 with a pronounced spinal curvature, apparently suffered during shipment (Bernstein, 2012). With the use of muscle relaxants, anti-inflammatory medications, laser therapy and chiropractic help, Khalid made a complete recovery.

Legs

When George Buffon saw his first giraffe, he later reported that the front legs were twice as long as the back. The front legs are slightly longer unlike in most ruminants, but appear to be much more so because of the sloping back which leads into the long neck (Colbert, 1938). The leg bones of giraffe are highly specialized, with the first, second and fifth metapodials absent. The ulna is not well developed. The hooves are extremely hard and flat, with the fetlock nearly touching the ground; this configuration along with large footpads give the giraffe good support for its large weight. The front hoof is slightly wider than the hind hoof, probably because of the greater weight on the forelegs.

Digestive system

Compared to large grazing ruminants, the giraffe has less-developed masseter muscles, larger mandibular glands, teeth with less chewing efficiency, a smaller rumen and a smaller omasum (Pérez *et al.*, 2012). These differences are related to its diet of leaves, rather than grasses, which is why giraffe have problems with zoo diets comprised largely of hay and concentrates. (See Chapter 10.)

Mouth

The large, 45-cm long prehensile tongue is the star of the giraffe's mouth. It is a wonder: almost black (perhaps as a protection against the sun), very long and covered by papillae thought to be as numerous, and the tongue itself as powerful, as that of any mammal

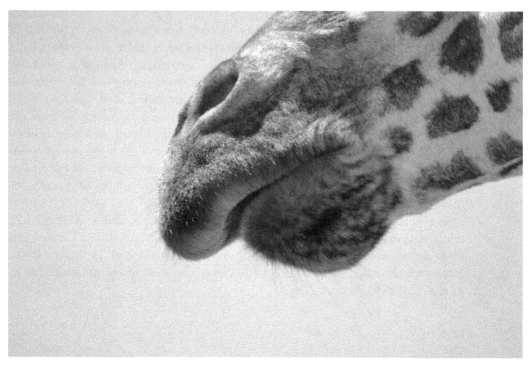

Figure 7.4   Giraffe have extra hairs around the mouth for protection against plant spines and thorns. Photo by Zoe Muller.

(Sonntag, 1922). (A giraffe at the Dortmund Zoo was born with a flesh-coloured tongue, but this may have been caused by inbreeding; Sicks, 2009.)

Skeletal muscle fibres forming the bulk of this organ are grouped into bundles arranged in different directions to allow great mobility (Lahkar *et al.*, 1992). When a giraffe throws back its head to browse high, the tongue can stretch straight up, with no upper incisor teeth to impede this. The transverse muscles contract when it reaches out to obtain a high morsel, while the longitudinals contract to bring the tongue back into its elongated mouth, just as these muscles do in other ruminants. A giraffe's flexible tongue can delicately remove the leaves of a branch it is already holding in its mouth (Dagg, 1970b). Or wrap its tongue around a branch and draw it gently into its extended lips. Or grasp leaves even among thorns with its prehensile lips (covered with thick hairs to protect them against the thorns) and pull. As in many other mammals, the tongue is covered by a mucous membrane lined with keratinized stratified squamous epithelium which varies in thickness in different areas but is thickest on the top (Lahkar *et al.*, 1989).

The mucous membrane on the free portion of the tongue is modified to form different types of papillae which are varied in structure and number over the three main areas: the flexible free apex at the front, the wider mid-area part of the tongue attached to the mouth floor and the root (Sonntag, 1922; Lahkar *et al.*, 1989).These papillae* can be large. They comprise:

Figure 7.5   Young Rothschild's giraffe reaches her tongue into her nasal cavity. Note thinness of neck. Photo by Zoe Muller.

- filiform papillae which are concentrated over the dorsal apex part of the tongue and to a lesser extent on the mid-tongue; many of the 50 or so projections have spines directed backwards to help in gripping leaves;
- fungiform papillae which are club-shaped, abundant and scattered around the entire dorsal surface of the tongue, more numerous at the apex than further back. These contain taste buds;
- lenticular papillae which have flat or convex surfaces present on the top of the tongue;
- conical papillae which run along either side and are either conical in shape or flat and broad; and
- circumvallate papillae which are convex in shape, varied in size and arranged along each side of the root of the tongue in two regular rows. The papillae at the root of the tongue, which also contains taste buds, are especially large; in this posterior area there is a thick layer of mucous glands.

The hard palate extending back to form the roof of the mouth has 18 transverse folds, their free edges turned backwards, extending back to the level of premolar 3 (Pérez *et al.*, 2012). In general, except for its more elongate shape, the mouth of the giraffe is similar anatomically to that of the cow, with glands as follows (Kayanja, 1973):

- small parotid salivary glands have two lobes, one rostral and one caudal, with tubes which open into the mouth in front of the upper premolar 2 (Kayanja and Scholz, 1974);
- mandibular salivary glands have two lobes each;
- sublingual salivary glands comprise monostomatic and polistomatic forms; and
- dorsal, ventral and intermediate buccal salivary glands are large, as they are in all ruminants (Kayanja, 1973).

Small palatine tonsils about 5 cm × 2.5 cm lie at the lower back of the throat (Hett, 1928).

Once food is in the mouth, it is directed to the central oral cavity between the molars by papillae (some up to 1.6 cm long) and callous processes lining the walls (Schneider, 1951). The premolars and molars have plicated layers of enamel running longitudinally with the jaw line so that food is chewed with a sideways motion of the lower jaw.

Stomachs

The oesophagus is highly muscular, allowing a giraffe to swallow food and regurgitate boluses quickly. It leads to the first stomach, the capacious rumen*, where food is mixed with gastric juices and made into boluses for later rumination as discussed in Chapter 5. The rumen with its thick epithelium and large papillae (Hofmann, 1968) has a dorsal and a ventral sac, as well as two blind caudal sacs. Hall-Martin and W. D. Basson (1975) made monthly analyses of the crude protein, fibre, fat, ash and gross energy contents in the rumen of dead wild giraffe, while W. van Hoven and E. A. Boomker (1985) report concentrations of hydrogen, carbon dioxide and methane in the rumen of other wild giraffe. C. J. Kleynhans and van Hoven (1976), who analysed the ciliate population from the rumen of 8 Cape giraffe, found 26 species of protozoa including 2 new to science. The most common genus was *Entodinium*.

The second stomach, the small reticulum, contains water swallowed by the giraffe (Hofmann, 1968). When a bolus is ingested after rumination, it goes not to the rumen but to the omasum or psalterium, the third stomach. (In nursing young, milk goes here directly, bypassing the first two stomachs.) Again the food is kneaded, this time against alternating rows of broad and narrow cornified lamellae lined with short papillae. Here superfluous alkaline liquid is absorbed before the food mass passes into the fourth stomach, the abomasum, which is the organ most like the stomach of non-ruminants. It has a thin lining with fine villi scattered over about 25 longitudinal wavy folds, again with digestive juices secreted from the walls which mix with the food. The opening to the duodenum is guarded by the strong muscular pylorus.

Intestines

The intestinal tract where the now-liquified 'food' is absorbed into the bloodstream is about 10 cm in circumference and comprises the typical small intestine (divided into duodenum, jejunum and ileum), caecum and large intestine. For adult giraffe all these components in total average 76 m in length (see chart of measurements for 15 individuals in Dagg and Foster, 1982); the small intestine itself is about 49 m long and the cylindrical caecum (which is similarly small in all artiodactyls with complex stomachs) about 0.65 m with small glands opening into its lumen (Derscheid and Neuville, 1925). When the food mass has passed through the small intestine, digestion has been completed. In the 25-m long large intestine, water is absorbed as the colon progresses by four distinct loops to join the rectum and the anus, from which faecal pellets shaped like acorns are excreted in bunches.

Gastrointestinal blockages are rare in giraffe, but 3 cases of colonic obstruction in captive reticulated giraffe have been reported over 16 years to the American Zoo and Aquarium Association, all of which resulted in death (Davis *et al.*, 2009). Symptoms include the animal losing weight and producing limited faeces, but unless there is early surgical intervention, the prognosis for affected animals is poor. Prognoses for an obstruction in the ileum are also poor (Gardner *et al.*, 1986).

Chronic diarrhoea may be a sign of exocrine pancreatic insufficiency-like syndrome in giraffe, as noted in four animals from the Warsaw Zoo (Lechowski *et al.*, 1991). The giraffe defecated six or seven times a day, discharging pulpy greenish, smelly excrement. One symptom was a decrease of amylase and lipase activity in fresh faeces compared to that present in the faeces of healthy giraffe. The sick giraffe were given pancreatic enzyme supplements which cured two of them, but the others died from acute rumen distension.

The digestive system of the giraffe is of special interest because it is the largest ruminant and a strict browser. Its anatomy, therefore, could be adapted primarily for its size or for its choice of diet. W. Pérez and colleagues (2009) describe in detail the gross anatomy of the giraffe intestines, then speculate on why the small intestine is relatively short in the giraffe and the large intestine relatively long, unlike in grazing ruminants where the relative proportions are reversed. They note that cattle, which have been studied intensively for many years, have faeces in pies rather than pellets (and are therefore less conserving of moisture); might have more microbial protein from the forestomach (needing time in the small intestine to be digested); and might have a small large intestine because of competition for space in the crowded abdomen. Or it may be that cattle themselves don't represent typical grazing ruminants, and giraffe should be compared instead with other ruminant species that have not yet been studied in detail.

Liver and gall bladder

These internal organs (along with others) were of great interest to early anatomists (see 12 references in Dagg and Foster, 1982). The large liver which has only a trace of lobulation is a key organ in the metabolism of food stuffs after they have been digested, and a storage place for carbohydrates in the form of glycogen. It secretes bile into the gall bladder, which empties via the choledochal or common bile duct into the duodenum 15 or 20 m past the pylorus where it helps in the digestive process. The right and left hepatic ducts and sometimes the thin pancreatic duct also empty into this common duct.

Anatomist A. J. Cave (1950) who dissected 19 giraffe recorded that only 2 of them had a gall bladder – one an abnormal double organ and the other (in a foetus) a tiny structure which he felt was in regression and would have disappeared had the foetus lived. Later researchers, however, disagreed with his findings: two adult females did have gall bladders (one over 11 cm long) and one adult male had one 11 cm × 5 cm (Kobara and Kamiya, 1965; Wakuri and Hori, 1970).

Nervous system

Recently, a coalition of researchers from South Africa and Sweden have banded together to describe in detail the central nervous system and brain* anatomy of the giraffe, with particular interest in whether they resemble the same organs found in other artiodactyls.

Ludo Badlangana and colleagues (2007a) have done a thorough study of both the gross and the microanatomy of the corticospinal tract of three male giraffe. Does it differ in any significant way from that of other ungulates because of the giraffe's long neck? This tract of the spinal cord contains axons that travel from the cerebral cortex of the brain to various levels of the spinal cord, its chief function being to mediate voluntary movement. The spinal cord, which can measure 2.6 m in a large male, extends to the sacrum, which is typical for artiodactyls. The primary motor cortex (M1) architecture closely resembles that found in sheep. Microscopic sections taken at three levels of the spinal cord indicate that the chemoarchitecture appears similar to those of other mammals, as does the pathway of the corticospinal tract through the brain. The researchers, who describe microanatomical features in depth, conclude from their work that the giraffe neck may be long, but the main features of its central nervous system are very similar to those of other artiodactyls.

A second paper by this group (Badlangana *et al.*, 2007b) describes the distribution and morphology of neurons in the medulla oblongata of a subadult giraffe. Within the medulla they found evidence for five putative catecholaminergic (neurons containing tyrosine hydroxylase) and five serotonergic nuclei. These two systems were, again, 'strikingly similar' to those present in other artiodactyls.

Three years later, some members of this group published their findings on the cholinergic, putative catecholaminergic and serotonergic nuclei in the diencephalon, midbrain and pons of subadult giraffe (Bux *et al.*, 2010). Again, the nuclear organization of these three systems was similar to those of other artiodactyls.

The organization and number of orexinergic neurons in the hypothalamus has been studied in the giraffe, and in the harbour porpoise for comparison (Dell *et al.*, 2012).

The sympathetic innervation of the cardiovascular system of the giraffe has also been described (Nilsson *et al.*, 1988).

Of evolutionary interest is the recurrent laryngeal nerve which, by connecting the human brain to the larynx, allows us to speak (Harrison, 1981; Prothero, 2007). This nerve in mammals avoids the direct route from the brain to the throat, instead looping around the aorta near the heart* before reaching the larynx. As the giraffe evolved its long neck, with it came a super-long laryngeal nerve which descends in the neck to loop around the aorta, then re-ascends to reach the larynx. It requires 15 feet to cover an effective distance of 1 foot. This design is not only wasteful, but one which is more susceptible to injury. Evolution-deniers are sometimes presented with this odd anatomical fact, which may or may not change their minds.

Kidney

As in other ruminants, the right kidney of the giraffe is situated near the diaphragm and the left placed farther back. In his study of the anatomy of this organ, N. S. R. Maluf (2002) describes the intricate parts in detail, noting the striking feature that the extensions of the renal pelvis reach deep into the outer medulla. Maluf postulates that urine within these extensions may diffuse into the medullary processes and from there be carried to the rumen, where microorganisms hydrolyse the urea to ammonia and thence form amino acids and protein, as occurs in sheep and camels. During times of drought, urea may perhaps be deposited into the renal interstitial tissue without the presence of ADH (antidiuretic hormone) which performs this function in human beings; there it may enhance the formation of concentrated urine for excretion in order to conserve water. Earlier studies of the kidney include that by K. H. Wrobel (1965), while J. Epelu-Opio (1975) has reported on the microscopic anatomy and ultrastructure of the urinary bladder epithelium.

Major muscles and ligaments

The various muscles of the giraffe neck (longissimus, thoracic and cervical, spinalis and semispinalis, cranial and caudal head oblique, and the multifidus muscles) have all been modified during evolution (Endo *et al.*, 1997). Their tendons are elongated, as are their muscle bellies. These muscles which occupy the spaces between the vertebrae support the head and neck by means of their wide attachments to the spinous and transverse processes of the neck bones. The nuchal ligament*, made up of thick elastic fibres, is particularly strong, supporting as it does the neck and head.

The masseter muscle which raises the lower jaw is the largest muscle of mastication in the giraffe, used for grinding solid plant material (Sasaki *et al.*, 2001). Unlike many smaller ungulates, giraffe do not have especially strong or complex masseter muscles, presumably because with their long necks and tongues they can reach soft, high-quality browse that does not require extensive chewing. In any case, this muscle in ruminants is less developed than in perissodactyls such as the horse because ruminants chew their food twice, the second time as half-digested cud.

The plantar flexors (plantaris, medial and lateral gastrocnemius muscles) of giraffe feet have similar fibre types during postnatal growth as are present in other mammals (Roy *et al.*, 1988). At birth, however, the giraffe has an unusually high percentage of light ATPase (adenosine triphosphatase) fibres near the bone attachments of these muscles. This indicates a precocial fibre type of development in giraffe which may be related, in part, to its large size and also because the newborn calf is able to stand almost immediately after birth. The plantaris muscle in the giraffe has apparently assumed the weight-supporting role normally performed by the soleus (the longitudinal muscle at the back of the lower leg), which is absent in this species.

Because its skin must fit tightly around its body for blood pressure control, the giraffe has no cuticular muscles to enable it to twitch its skin to discourage insects (Bush, 2003).

Non-digestive glands

Giraffe do not have preorbital glands, inguinal glands or interdigital glands because they do not communicate with each other by scent (Pocock, 1910). As well as the buccal or salivary glands mentioned earlier, they do have:

- thyroid and parathyroid glands in the neck which secrete hormones (Frank, 1960);
- adrenal glands (near the kidneys) which secrete adrenaline and noradrenaline and are slightly larger than the thyroid glands (Crile, 1941). These glands were examined under an electron microscope and described after they had been removed from six pregnant giraffe who had been chased for 30 min before being shot (Weyrauch, 1974);
- a pituitary gland underneath the brain which is relatively much larger than that of an elephant and which secretes a number of important hormones (Hanström, 1953; Kladetzky, 1954); and
- active lacrymal glands, perhaps related to the large size of their eyes – at the Taronga Zoo some individuals had tears which dropped onto the ground.

8 Physiology

Organs marked with an asterisk * can be viewed in colour on the website given near the beginning of Chapter 7.

Immobilization

If you want to find out how the various systems of giraffe work, unfortunately you have to immobilize one to carry out the necessary procedures – an unpleasant experience for the animal. This process in the wild must be used, for example, to put a collar on an individual for ecological studies, to collect semen or blood or to restrain it for more invasive procedures involving blood pressure and respiratory research.

Vaughan Langman (1973) describes the capture and immobilization of 21 giraffe (19 survived) in South Africa using a combination of fentanyl citrate, azaperone and hyoscine hydrobromide. Nalorphine reversed their reactions and returned the animals to their normal state. Mitchell Bush and Valerius de Vos (1987) over 2 years immobilized 15 giraffe (of which 13 survived) in the Kruger National Park. In the process they refined their use of drugs, in this case employing carfentanil and xylazine (with a variety of reversal drugs) to ensure the best case scenario. Possible disasters arising from anaesthesia include tachycardia, hypotension, cardiac failure or hyperthermia. Although the process may be considered a success if the giraffe survives to walk away to freedom, the animal may actually be in jeopardy for the next week, dying from causes such as radial paralysis, aspiration pneumonia or pulmonary excitement (Langman, 1973).

These teams of eight or so people offer advice on the best way to go about immobilization in the wild. First, it is important to choose carefully which giraffe to immobilize, preferably in the early morning when it is not too hot. It should be an animal:

- whose weight can be accurately judged so that the correct dosage of drugs can be calculated;
- who is part of a group and therefore calmer than if alone;
- who is not near water and therefore probably has not drunk recently;
- who is on flat terrain without gullies, holes or bushes so it can be followed easily by the research team; once an animal is narcotized, it becomes oblivious to its surroundings and cannot be herded by a vehicle; and
- who tends to be smaller rather than larger and therefore more easily manipulated.

Figure 8.1   Rothschild's giraffe darted with anaesthetic for capture. Photo by Zoe Muller.

After firing a projectile dart containing an anaesthetic from a modified shotgun into the flank or neck of the chosen wild giraffe, other problems must be addressed:

- the dosage of drug has already been selected, but its efficacy depends on pre-darting excitement, amount of running of the giraffe and placement of the dart;
- in staggering about, the darted animal may fall over a log and land heavily on its side. The pressure of the fall on its abdomen may force contents of the rumen up its oesophagus into its mouth from where food bits may enter its trachea and lungs, causing inhalation pneumonia. One large male darted by Bush and de Vos (1987) died 40 min later from asphyxiation;
- a high dose of anaesthetic should be used to hasten its lateral recumbency (with legs on one side of the body) so that it cannot wander far away before collapsing (one darted animal annoyed its harassers by travelling 8 km before keeling over);
- however, if too great a dosage of a drug such as xylazine is used, it can predispose the giraffe to predation after its recovery;
- some animals develop marked respiratory depression and apnoea after darting which require immediate injection of the reversal drugs to save them;
- the recumbent giraffe is blindfolded to decrease visual stimulation, and there should be no loud noises or talking near it;
- the head must be kept always high and the neck well supported along its length, with the nose pointing down in case of vomiting. With sensation lost in the pharynx and

larynx (as well as elsewhere in the body) with a general anaesthetic, a tracheal intubation is useful to prevent aspiration of rumen contents that can cause obstruction or pneumonia. Intubation also allows the control of respiration, which may need to be increased; and

- a muscle relaxant such as i.v. glycerol guaiacolate prevents muscle spasms in the neck and legs.

A darted giraffe usually collapses unassisted, but if necessary it can be cast by two people holding a rope at shoulder height in front of the animal which is then lowered, crossed behind it and brought back to the front, confining the legs and bringing the animal to the ground. The head and neck must then be immediately held up to prevent regurgitation of food.

After the necessary procedures have been completed, reversal drugs are injected into the giraffe whose head continues to be held high by two people to keep it still for as long as possible before it is completely awake. Then a thick rope placed around the giraffe's shoulders and people pushing upward can help the giraffe rise to its feet from a position of sternal recumbency.

Langman reports various data for his non-drugged and drugged animals. The respiration rate per minute was faster in the drugged animals (means 23 and 19.8), as was the heart rate (means 113 and 81), but the body temperatures were the same (38.5°C). For four young animals who died after being immobilized, there was a period of neurotic excitement when they walked or ran about frenetically, their body temperatures rising, ranging from 39.4 to 42.3°C. Then they collapsed in exhaustion and death. Autopsies showed that two of the giraffe died from aspiration of rumen contents and the other two had the right ventricle of their heart distended.

More recently, a team in Tanzania has darted and immobilized 12 free-ranging adult giraffe in Ruaha National Park using 18 mg etorphine (Mpanduji *et al.*, 2010). For males versus females, respectively, the time from darting to first noticeable effect of the drug averaged 8 min for males and 12 min for females; the running time after darting averaged 15 and 17 min; the distances covered after injection time to recumbency averaged 1.3 and 2.2 km; and the total down time averaged 12.6 and 11.5 min. Three giraffe died during the exercise from regurgitation of food and general weakness, from failure of respiration because of poor positioning and from failure to recover from immobilization after entering thick vegetation. The team compared a number of biochemical parameters for the male and female giraffe.

In a semi-free-ranging environment such as the 380 acres of giraffe pasture in the Fossil Rim Wildlife Center in Texas, immobilization can be improved and simplified by positioning a padded board under the downed giraffe's neck and laying it at an angle of about 50° against one of the vehicles which surround the animal (Peterson *et al.*, 2010).

Anaesthetizing a giraffe in a zoo building is much easier and safer (but not entirely safe: one large bull suffered a spinal cord injury during recovery from anaesthesia; Aprea *et al.*, 2011):

- to prevent ingesta from the rumen possibly entering the lungs, the target animal is denied food and water for 48–72 h to minimize vomiting;

Figure 8.2   This Rothschild's male had collapsed after being darted; he was given an antidote 25 s before this photo was taken. He was transported to the original Rothschild's home range (see Chapter 11). Photo by Zoe Muller.

- after darting, keepers surround the animal to ensure that it subsides gently with its head always above the level of its rumen and its nose pointing down;
- drug dosage can be lower, as captive animals require less immobilizing drugs than free-living ones; and
- the giraffe can be kept away from objects in the paddock which could harm it (whereas holes, gullies and fallen trees are a problem for narcotized giraffe in the wild).

Nadine Lamberski and Andy Blue (2010) describe all the precautions that a zoo should take before anaesthetizing a giraffe.

The treatment of young calves may be different. Dennis Geiser, Patrick Morris and Henry Adair (1992) operated on a one-month-old reticulated male giraffe who had suffered a fractured left metatarsus. They had to anaesthetize the youngster a number of times during his extensive treatment which enabled them to determine how this could be done most effectively. He tended to hypoventilate during anaesthesia, resulting in respiratory acidosis, but this could be remedied by monitoring pulse oximetry and end tidal carbon dioxide. They advise that tracheal intubation and assisted ventilation be routinely used with anaesthesia in young giraffe.

Mitchell Bush (Bush *et al.*, 2001; Bush, 2003) gives a compendium of chemical restraint agents for sedation of giraffe, along with haematological and serum biochemical parameters. More recently, he and Scott Citino (Citino *et al.*, 2006; Bush, 2010) recommend an anaesthetic protocol using thiafentanil, medetomidine and ketamine. Citino (2008) has also devised a simple ventilator designed for emergency use for giraffe captured in the wild. It is based on an electric or gas-powered leaf blower which was demonstrated to be effective for three giraffe.

Cardiovascular system

When I saw my first wild giraffe, she was drinking at a small pond near the road along which we had been driving. As I stared at her in awe, she whipped her head into the air, water droplets flung about, and turned briskly to walk back into the bush. I was astounded. I had been waiting all my life to see a giraffe in the wild, and had assured myself that she would have had to raise her head gradually because the blood pressure change in her brain would be great – from ground level to 14 feet in the air. So much for armchair zoology.

Because of the height of a giraffe, there is great interest in its cardiovascular system. 'Wow', a cardiologist said to me about my interest in giraffe. 'They must have a huge blood pressure', which they do, about 200 mmHg at heart level, as well as many other adaptations; otherwise, as a later researcher postulated, judging by human standards a giraffe lifting its head after drinking water would faint from lack of blood to the brain. Distal parts of the arterial system are exposed to variations in blood pressure of more than 400 mmHg during their daily lives, while proximal carotids near the heart are chronically distended by pressures of over 300 mmHg (van Citters *et al.*, 1968). These pressure ranges and their corresponding wall tensions are greater than those present in any other animal.

How does blood find its way 2.5 m to an adult giraffe's head, then keep the giraffe from blacking out when it reaches down 5 m to drink water, or again when it whips its head back up? Why doesn't blood pool at the end of the long legs when there is little obvious pressure to push it back up to the heart?

Heart

For many years the heart of the giraffe was thought to be much larger proportionately than that of other mammals, which is how it was presumed to be able to develop the necessary high blood pressures. However, the data of R. H. Goetz, who measured the heart of a 500-kg young giraffe as 11.25 kg, or 2.5% of the body mass (and five times that predicted for other mammals), was incorrect (Goetz, 1955; Goetz and Keen, 1957).[1]

Graham Mitchell and John Skinner (2009) collected anatomical data related to the heart and its growth from 56 giraffe of all ages, including 10 foetuses. These had been newly killed in Zimbabwe between 2006 and 2009. After the removal of all fat tissue, the

---

[1] The authors note that 'No institutional animal care approval was required for this study. Animals were culled according to the legal requirements of Zimbabwe.'

detachment of the large blood vessels and the draining of blood, the largest heart mass was 7.8 kg belonging to a 1441-kg old male. The relative heart mass declined somewhat with body growth, similar to that in other mammals, but the thickness of the walls of the left ventricle increased markedly with growth; the maximum outer wall was 8 cm thick and the interventricular wall 5.7 cm thick, much thicker than those of other mammals. This hypertrophy correlates with increases in the length of the neck and with hypertrophy of the tunica media of arterioles. The data suggest that the cardiac output of giraffe is similar to that of other mammals, but that resistance to blood flow in both the systemic and pulmonary circulations is higher, as therefore is the blood pressure. The heart of the giraffe is not proportionately larger than that of other mammals, but its thick left ventricle walls indicate that it is functionally very efficient.

The blood pressure at the heart is related to the slope of the neck (Mitchell and Skinner, 2009). If the neck is 55° to the vertical in an adult (775 kg) giraffe, the blood pressure is 208 mmHg, nearly twice that of mammals of equal mass. At angles of the neck to the vertical of 30°, 60° and 90°, the blood pressures, respectively, were calculated as 187, 216 and 247 mmHg. The blood pressure at the brain of giraffe and all other mammals including ourselves is about 100 mmHg, that required to achieve adequate cerebral perfusion. The pulmonary blood pressure in an adult giraffe, that in the pulmonary artery from the right ventricle to the lungs, is about 38 mmHg compared to 15 mmHg in other mammals.

From their extensive data, the researchers conclude that as a young giraffe grows, the gravitational pressure of the arterial column of blood in the neck stimulates increased muscle build-up in the ventricle walls; therefore, the physiological response to the elongating neck is cardiac hypertrophy, not cardiac enlargement or an increase in cardiac output. At the same time, arteries and arterioles at or below the level of the heart also hypertrophy, reflecting high arterial pressure, high gravitational pressure and the effects of autonomic regulation of resistance. Increased blood pressure results from a balance between cardiac hypertrophy and increased arteriole resistance, a resistance that prevents tissues from being subject to extraordinarily high blood pressures. The heart of a giraffe is relatively small but powerful and well adapted to the giraffe's unusual anatomy; its internal and external anatomy has recently been described in detail (Pérez *et al.*, 2009).

Blood vessels

The distance between the heart of an adult giraffe and its brain can be 250 cm (about 8 feet) (Maluf, 2002). A column of blood that height equals a 196-mm column of mercury. Mean arterial pressure + perfusion pressure must exceed this gravitational hydrostatic pressure if blood is to reach the brain at a pressure of 100 mmHg. Because the giraffe has the longest neck of any mammal, it also has the highest normal arterial blood pressure of any known mammal. The mechanism of its hypertension is unknown, and hence called 'primary' or 'essential'.

The cardiovascular anatomy of a foetal giraffe is described by Keen and Goetz (1957). The carotid arteries in the neck, which have to withstand elevated shearing forces because of the high blood pressure, are largely muscular in the adult, but in the foetus they have much more elastic tissue (Kimani, 1983a). As the young animal grows, there is an eightfold increase in the cross-sectional area of the carotid involving the development of smooth muscle bundles in the outer medial thickness of the artery walls. The walls also have collagen fibres linked to the smooth muscle cells (Kimani, 1981a). This linkage may facilitate transmission of the force of contraction between the cells and to the surrounding connective tissue framework.

Microscopic analysis further shows that the innermost layer of the carotid arterial vessels which deliver blood from the heart to the head, called the tunica intima, contains a prominent subendothelial elastic layer which diminishes in size from the lower neck to the head, in keeping with the blood pressure profile of the giraffe (Kimani, 1983b). J. K. Kimani and colleagues have carried out extensive further research on the carotid arterial system in the giraffe (Kimani, 1981b; Kimani and Mungai, 1983; Kimani and Opole, 1991), on the vertebral artery (Kimani, 1987) and on intracranial arteries (Kimani et al., 1991a). The arterial vascularization in the brain of the giraffe is also described by Frackowiak and Jakubowski (2008).

The diameter of the coronary arteries which supply blood to the heart muscle have a nearly linear correlation to the virtual heart diameter (calculated from the cubic root of heart weight) in mammals ranging in size from mice to giraffe (Thüroff et al., 1984). For the larger animals, however, the diameter of the arteries is somewhat larger. Their pulse rate, of course, is far slower than that of a mouse (500 beats/min).

The arteries below heart level have relatively more muscle than those above, and smaller lumens (Maluf, 2002). Because of this musculature, the diameter of the arterial wall of the femur is greater than that of the tibia, and much greater than that of the metatarsal. The arteries at heart level (coeliac, anterior mesenteric and renal) lack this extra muscle tissue, while the muscle layers even in arterioles at the level of the heart hypertrophy with age as a response to rising mean arterial blood pressure (Mitchell and Skinner, 2006). The large arteries in the legs may seem to be passive conduits of blood, but they have an important role in converting pulsing flow to a more continuous flow and thus influencing the workload of the heart (Østergaard et al., 2011).

Blood pressure

The circulation of the blood is managed by blood pressure differences which are complex. Arterial blood pressure just beyond the heart* of a giraffe is about 200 mmHg, twice that in human beings. It propels the blood upward through the carotid arteries until it reaches the brain where the pressure is about 100 mmHg, similar to that at the human brain. Arterial blood pressure is generated by cardiac output and peripheral resistance, both of which are regulated by baroreceptors. When blood pressure falls, baroreceptors are activated, causing cardiac output and peripheral resistance to increase, and vice versa.

The two external carotid vessels each divide, just before reaching the brain, into many small vessels which form a tight network called the *rete mirabile*. These vessels have elastic walls which expand to hold more blood when the head is lowered so that the brain is not flooded by this excess. A further safeguard for the brain while the head is lowered is a connection between the carotid artery and the vertebral artery which drains off some of the blood even before it reaches the *rete mirabile*. The connection ensures that the *rete* is always supplied with blood. This network is elastic enough when the head is raised quickly to retain sufficient blood so its supply is not momentarily depleted before the system has had time to readjust to the pressure changes (Lawrence and Rewell, 1948).

To understand blood pressure issues in Masai giraffe, researchers captured two bulls in the wild for a blood pressure study, quickly stitching a transducer into their right carotid artery so that blood pressure changes could be measured when they were freed (van Citters *et al.*, 1966; Warren, 1974). As they then walked about and browsed, their blood pressures ranged between 180/120 and 140/90. When one male galloped, its heart rate increased from under 60 to 170 beats/min with a peak blood pressure of 230/125. Later the same day, the giraffe were again darted and anaesthetized so that the instruments could be removed. After they had lain on the ground for 10 min, their carotid blood pressure was 280/160 and their heart rate 100 beats/min. They also survived this second operation, apparently unharmed and able to go on with their lives. (See Dagg and Foster, 1982 for other early experimental results.) As a comparison, for five giraffe immobilized at the Taronga Zoo in Australia, their mean respiratory rate, heart rate and rectal temperature were, respectively, 12.2 breaths/min, 58.6 beats/min and 37.1°C (Vogelnest and Ralph, 1997).

The blood pressure of a mammal depends on cardiac output and peripheral resistance. If there is an increase in blood pressure, it will be caused by an increase in one of these variables or in both of them. Similarly, if there is a decrease, it will be caused by a decrease in one or both of them. When the head of a giraffe is suddenly raised from drinking, the perfusion pressure of blood at the brain falls from about 200 mmHg to about 120 mmHg. This fall is too large to be compensated for by a reflex increase in heart rate and aortic blood pressure. T. A. McCalden and his colleagues (1977) found by experimenting with immobilized giraffe that there was no significant change of carotid blood flow from normal even when the brain perfusion pressure varied by as much as 70 to 210 mmHg. This relatively constant blood flow indicates that the resistance of the cranial circulation varies in order to minimize the effects of the pressure change. This range over which significant change in carotid flow would occur is much greater in the giraffe than in other species.

More recently, Graham Mitchell and colleagues (2006, 2008) addressed problems of blood pressure in giraffe by simulating the circulation of liquid in a mechanical model of a giraffe neck and head. They concluded that the mean arterial blood pressure is a consequence of the hydrostatic pressure generated by the column of blood in the neck and that no special effect (as had been hypothesized) assists cranial circulation. They also studied with this model the effects of head-raising and head-lowering in giraffe which allowed them to estimate cranial pressure changes during these postures. They concluded that an increase in cerebral perfusion pressure (which is the difference between the mean arterial pressure and intracranial pressure) is indeed significant in maintaining cerebral blood flow during head-raising. Mitchell along with two colleagues (Paton *et al.*, 2009) argue that 'Cushing's mechanism' helps control blood pressure; it is a physiological mechanism for long-term control of mean arterial pressure with short-term control the province of baroreceptors. The mechanism is present also in hypertensive rats and human beings with essential hypertension. Research on anaesthetized giraffe by E. Brøndum and 17 colleagues (2009) also found no support for a siphon mechanism in the circulation system, but does show that when the giraffe head is lowered, blood accumulating in the jugular vein* affects the mean arterial pressure.

Venous blood heading from the brain down to the heart via a jugular vein has a low blood pressure of about 16 mmHg near the brain and 7 mmHg near the heart (Hargens *et al.*, 1987). The veins have a series of valves opening downward when the giraffe is standing so that the column of blood is continuous, with gravitational pressure operating along its entire length (Badeer, 1997). When the animal bends down to drink water, however, these valves fill with venous blood, helping prevent an excess rise of blood pressure in the brain. Graham Mitchell, Sybrand van Sittert and John Skinner (2009b) who measured the jugular valves in 60 veins of 30 dead giraffe found there were about 6 or 7 in each vein of males, females and foetuses. Most (88%) were bicuspid in form and most valves were located nearer to the torso than to the head. The main function of the valves is to ensure that when a giraffe's head is at a level below the heart, blood returning from the body to the heart is diverted into the right atrium and does not enter the jugular vein. Therefore, cardiac output is maintained and when the head is lowered there is no change in arterial pressure. The amount of blood that accumulates in the jugular vein during head-lowering is a tiny fraction of total blood volume, so its accumulation does not affect the blood pressure. Further research using models to understand the functioning of the jugular vein is discussed by Hicks and Badeer (1989), Pedley *et al.* (1996) and Brook and Pedley (2002).

How does the giraffe prevent swelling in its legs, with blood and tissue fluid (oedema) pooling there rather than the venous blood returning expeditiously to the heart for recycling? This lower body circulatory system is the reverse of that to the head. It is the arterial blood leaving the heart for the torso and legs that is aided by gravity while the venous blood in the feet must somehow be forced upward.

The feet of a standing giraffe have an arterial pressure of about 260 mmHg, the sum of the hydrostatic pressure from a blood column of about 2 m (the distance between heart and hoof) and the mean arterial pressure (Pedley, 1987). The venous pressure comprises about 150 mmHg (Millard *et al.*, 1987) and the tissue pressure about 44 mmHg (Hargens *et al.*, 1987). (By contrast, pressure values in the feet of moving giraffe vary hugely: arterial pressure between 70 and 380 mmHg, venous pressure between −250 and +240 mmHg and tissue pressure between −120 and +80 mmHg.)

If a giraffe is walking or running, muscle contractions in the legs compress the veins against bones, forcing blood upwards. Giraffe standing still are more of an enigma, with several ways of propelling venous blood in the feet back to the heart against the force of gravity (Hargens *et al.*, 1987). The bodies of giraffe are encased in a tight fascia skin layer (anti-gravity body suit) that ensures inflexibility (except at joints) and prevents the pooling of venous blood (moisture) in the feet and therefore the development of oedema. As well, one-way valves in the veins and lymphatics help propel blood and lymphatics upwards; special 'seried' valves are present at the entry of major tributaries into the axillary and brachial veins (Rewell, 1987).

The amount of blood flowing to the lower limbs is controlled by adaptations to the arterial blood vessels, with the lumina of leg arteries much reduced. The posterior tibial artery of a giraffe and the carotid arterial artery near its head both have an outside diameter of 3 mm, but the wall thickness of the former is 7 times great than that of the latter, and the lumen 6 times smaller (Paton *et al.*, 2009). Blood flow through both vessels is the same, but the pressure in the tibial artery is 4–5 times greater. Arterial blood pressure is not transferred to the veins because the arterioles are separated from the veins by sphincters and an extensive capillary bed. Other characteristics of the blood pressure system are:

(a) arterial wall hypertrophy and precapillary vasoconstriction also reduce dependent blood flow;
(b) low-permeability capillary membranes retain intravascular proteins, preventing oedema;
(c) a prominent lymphatic system is in effect outside the blood circulatory system; and
(d) in the neonate, a leg artery is already narrow beneath the knee, so that this feature is a preadaptation rather than a result of blood pressure changes after birth (Østergaard *et al.*, 2011).

Respiratory system

Because of its long neck, the giraffe's respiratory system would seem to be unique. In the past, researchers have hypothesized that the long trachea*, comprising 200 cartilaginous tracheal rings (Cano and Pérez, 2009), must act as a large dead space between the nasal intake of air and the lungs (Harrison, 1980). Two studies have claimed that respiration is slow either because of a high tracheal resistance or because of an abnormally large dead space; another study claims that respiration is fast because the animal has to hyper-ventilate to compensate for the large dead space (Langman *et al.*, 1982). Obviously it has been difficult to determine a giraffe's normal rate of breathing because if stressed by measuring equipment, its breathing accelerates. However, it is possible to count respira-tory movements in a free-living giraffe just by patient observation (Graham Mitchell, pers. comm., January 2013); the normal respiratory rate is around 8–12 breaths/min depending on age and size.

Vaughan Langman, O. S. Bamford and Geoffrey Maloiy (1982) decided in their research to study two main questions: does the trachea of a giraffe have a large dead space, and if so, does this affect pulmonary ventilation? To do this they worked with three Masai giraffe, two females and a male, who had been held for 2 years in an enclosure near where they had been captured. After careful training, the giraffe were able to take part in a respiration experiment without panicking, a notable achievement (Langman, pers. comm., December 2012). This involved being held in a crush and wearing a mask over their heads so that air in and out of their lungs could be measured and collected for analysis (Figure 8.3). The ambient and their rectal temperatures were also taken at each session. This apparatus enabled the men to document for each giraffe its tidal volume, ventilatory minute volume, oxygen consumption, and respiratory water loss and utilization. In addition, for the sake of comparison, necropsies were performed on 88 individuals of 16 species (31 of them giraffe) from the wild so their respiratory organs could be dissected.

The researchers found that although the trachea was long, it was also narrow. The relationship between body weight and tracheal volume was similar to that found in other large mammals, so the trachea did not have an unusually large dead space after all. The observed values for all the respiratory variables were also similar to those of other 600-kg mammals, so the giraffe does not have a unique respiratory system despite its unusual shape. Its respiration rate is neither very high nor very low, but an average of nine breaths per minute (which varies with ambient temperature). (Earlier values were high because of the stress animals endured during tests – some blindfolded giraffe were tied to a scaffold and in one case a giraffe died.)

Lungs

As in most ruminants, the left lung of the giraffe has two lobes and the right three main lobes plus an extra lobe. Lobular divisions of the lung and bronchial ramifications are described by S. Nakakuki (1983), and hydatid cysts on giraffe lungs by Bozkurt *et al.* (2011).

In a physiological study, Graham Mitchell and John Skinner (2011) carried out extensive measurements on the lungs of 46 giraffe, all newly killed, ranging in age from a 5-month female of 147 kg to a male of over 20 years of 1441 kg. There was no significant difference between the values of the measurements they obtained for males and females of similar ages, so they pooled the data for the various age groups.

The range of lung mass for the giraffe was 1.2 kg (for the youngest) to 9.0 kg (for the largest), which is significantly lower than the predicted mass of 1.6–15.2 kg for other species of similar sizes. The lung volume of giraffe, therefore, was also smaller than predicted from other species, ranging from 7.7 to 69.3 litres; other species of similar size had ranges of 10–112 litres. In almost all mammals the lung volume increases as an animal grows, but this is not true of giraffe, whose lung volume decreases relative to increases in body mass. Presumably this is because the bulky skeletal structure to support the heavy neck increases too, leaving little space in the abdomino-thoracic cavity which holds the digestive tract and other organs as well as the lungs.

The lungs' tidal volume (volume of air inhaled and exhaled at each breath) is larger than in other big animals – about 6 litres for a 600-kg giraffe – even though the trachea narrows with growth which helps to limit the increase in dead space volume (Langman *et al.*, 1982). The small lung volume and 2-m long trachea mean that giraffe have a reduced inspiratory reserve volume (about 6% of the air in the lungs compared to about 16% for most mammals) and high resistance to air flow. These limit the ability of giraffe to increase tidal volume and prolong the time it takes for them to inhale and exhale. They breathe slowly, 8–10 times/min, presumably because of the small diameter of the trachea related to its length; by comparison, camels breathe about 13–14 times/min and people 14–16 times/min (Hugh-Jones *et al.*, 1978).

Figure 8.3   Vaughan Langman holds the head of a male Masai giraffe in order to record the temperature of his exhaled air. The animal was trained to cooperate in only 4 months. Photo by Vaughan Langman.

Figure 8.4   Knut Schmidt-Nielsen and Dick Taylor on the ground, while Vaughan Langman holds the Masai giraffe's head. Photo by Vaughan Langman.

Despite these apparent disadvantages for giraffe, their minute and alveolar ventilation rates are similar to those of other animals of the same size so the oxygen supply to tissues is not compromised (Mitchell and Skinner, 2011). Giraffe can outrun a horse, set up a steady pace for long periods and gallop at 60 km/h for at least 5 min (Langman *et al.*, 1982). If there is great oxygen demand, a giraffe increases its respiratory rate rather than its tidal volume. It cannot pant rapidly because of the flow resistance in the trachea (Hugh-Jones *et al.*, 1978). The giraffe is not restricted in its activities by its small lungs; oxygen consumption is determined more by the oxidative capacity of tissues than by the capacity of the lung for gas exchange.

Thermoregulation – coping in the heat

Giraffe, who often live in very hot dry climates, have a number of ways to keep cool.

Body shape

The larger an animal, the relatively less is its surface area. This is a disadvantage in hot climates because there is less body surface from which to dissipate heat. However, the giraffe's elongate shape gives it more surface area than that of a rhino, hippo or elephant with the same body mass; unlike giraffe, they resort to lolling in water or in mud wallows during hot days to keep cool. By the same token, in cold weather the giraffe's unusual shape makes it difficult for them to keep warm (Clauss *et al.*, 1999). (If giraffe are able to activate the muscular valves that exist in the upper leg arteries when they are cold as well as when they are exercising, they could ameliorate the temperature of their limbs; Kimani *et al.*, 1991a.) Perhaps members of the Masai and Watutsi tribes who live near the equator are also tall because of their hot climate (Schreider, 1950).

Body position

Recent studies show that if the temperature is very hot, as it can be in the Namib Desert, giraffe will seek shade or orient their bodies so that they are longitudinal to the sun's rays (Kuntsch and Nel, 1990). (Camels in the Sahara do this too; Gauthier-Pilters and Dagg, 1981.) Calves and females with their smaller mass seek shade more often than do adult males (Langman and Maloiy, 1989). By contrast, when the weather is cold, giraffe tend to stand perpendicular to the sun to expose a large body area to it (Kuntsch and Nel, 1990).

While studying giraffe in the 1950s in the eastern Transvaal province (now Mpumalanga) of South Africa, I tried to correlate giraffe behaviour with the weather (Innis, 1958). For example, I watched them on open slopes at various times and no matter how hot the sun or what orientation the wind, the animals faced a variety of directions. In the hottest part of the day (up to 40.5°C) they often lay down, however, and walked slowly if they moved at all. (Lions were rare in the area and many of the giraffe with drooping heads seemed to be asleep, so I am sure they were not facing different ways to be on the alert for predators.)

Seeking shade?

As a first line of defence, do giraffe seek shade as a hot day gets hotter? In the first six months of 1968, I carried out a temperature-related study of animals in the Taronga Zoo near which we were living during my husband's sabbatical in Sydney, Australia (Dagg, 1970a). The high daily temperatures over that period varied from 37 to 12°C. Every sunny day I pushed my young daughter in her stroller through the zoo, noting in each paddock whether the animals were in the shade or in the sun (Dagg, 1970a). The 18 giraffe preferred to be hot. On the 50 sunny days I visited their paddock, even when it was 37°C (99°F), most of the animals still stood or lay in the sun rather than seek the shelter of a building.

The body temperature was taken of various giraffe in Africa. On less than very hot days one young female had a body temperature of 37.75–39.1°C (Bligh and Harthoorn, 1965), while the body temperature of three adult giraffe averaged 36.7°C (Langman *et al.*, 1982). As we shall see, however, in very hot weather their body temperature rises to reduce water loss.

Succulent leaves

A 1000-kg giraffe eating 34 kg of acacia leaves containing 60% water takes in about 21 litres of preformed water a day. (Other plants may also have leaves with a fairly high water content, of course.) Metabolic intake of water is estimated at 3.5 litres per day, so the giraffe would have at its disposal a total of 24.5 ml/kg body mass (Taylor in Dagg and Foster, 1982; Mitchell and Skinner, 2004). Domestic cattle need far more water than giraffe each day and camels a little less. By eating succulent leaves, a giraffe usually does not need to drink water at all. However, if it is freely available, it may drink every day. (In a drought, with water scarce, giraffe could be in danger of developing stones in its urinary tract (urolithiasis) due to concentrated urine.)

Skin-covered horns

Heat may be dissipated through the ossicones on the head of giraffe which are covered in skin rich in blood vessels (although on the top of these 'horns' the hair is worn off in the large males because of their head-hitting workouts with each other). The arterial blood flows to the horns through the large branch of the cornual artery, a vessel seemingly too large for the needs of the ossicones themselves (Ganey *et al.*, 1990; Mitchell and Skinner, 2004). Then, theoretically, the cooled blood would flow into the venous drainage channels of the spongiform bone providing a layer of cool temperature between the brain and the outside air. However, if the ambient temperature is too high, this transfer of heat would go the wrong way and actually make the animal hotter.

Blood flow at the brain

The brain may also be cooled by the function of the carotid *rete mirabile*, which may or may not be able to lower its temperature below that of the rest of the body (Mitchell and

Skinner, 2004). The brain itself is fairly small – up to 1500 g, or about 0.1% of body mass (compared to humans at 2%), while the carotid *rete* tissue is relatively big, affording both adequate time and area for heat exchange when its blood comes into close contact with the cool venous blood near the brain.

Nasal cooling of the brain

Keeping cool in general is caused by body surface blood vessels dilating to allow convection and evaporation from the air. The air in the lungs of mammals is heated to body temperature and saturated with water vapour. When a giraffe breathes out, however, the exhaled air has a temperature substantially cooler than that of its body core (Figure 8.3). This happens because during inspiration, the nasal surfaces are cooled by the incoming air augmented by evaporation of water; during expiration, the warm air from the lungs is cooled by these surfaces giving up in the process heat and water. The amount of water recovered in this nasal exchange can reduce water loss in a giraffe by 1.5–3 litres a day. Langman and his colleagues (1979, 1982) note that the operation of this nasal heat exchange requires no metabolic energy even while it provides significant benefits; this factor would seem to favour its evolutionary development.

In the warm-blooded giraffe, the cooling occurs by evaporation taking place in the large nasal cavity which acts as a small refrigerator. This includes an elaborate turbinate architecture and a very large nasal mucosa surface area ideal for heat exchange (the air flow being only 2 mm from the mucosal surface) which conserves water and presumably also cools the brain via the carotid *rete* mechanism and the jugular venous blood (Mitchell and Skinner, 2004). Giraffe breathe almost exclusively through the nose, unlike people or dogs.

In experiments with Masai giraffe in the wild, Vaughan Langman and colleagues (1979) found that when the outside temperature was 17°C, the exhaled air was 13°C below core temperature. When the ambient air temperature was 24°C, the exhaled air was 7°C below core temperature. This phenomenon was not unique to giraffe as goats and waterbuck were equally as efficient in retaining water from their exhaled air. When the ambient temperature rises from 36 to 39°C, the respiration of a giraffe increases to about 12 breaths/min. This increases the volume of air moving through the turbinate architecture which can reach 130 litres/min (Mitchell and Skinner, 2004).

Raising body temperature

Giraffe inhabit hot regions of Africa. As Knut Schmidt-Nielsen (1959) had found for camels, they also have the ability to raise their body temperatures so that body water is not wasted by being used for cooling by evaporation, as when humans and other animals sweat or pant at high temperatures.[2] In their research, Langman and colleagues (1982)

---

[2] I was fortunate in the summer of 1973, along with my husband and daughter, to spend a day with Kurt Schmidt-Nielsen in France. He was the guest of his old friend Hilde Gauthier-Pilters who had helped him in his research on the physiology of camels. Hilde had worked in the past with Kurt who was well known

found that the body temperatures of giraffe ranged from 36.7°C on cool days to 39°C on torrid days. Values for oxygen consumption, respiratory frequency, minute volume and respiratory evaporative water loss were all strongly correlated with body temperature. Research on camels and eland have shown that their body temperatures also increase with increasing ambient temperature. Extrapolation indicates that if the same mechanisms work in giraffe, which seems likely, a small giraffe could save 4 litres of water a day in hot weather, and a large giraffe even more (Schmidt-Nielsen *et al.*, 1957; Taylor and Lyman, 1967; Dagg and Foster, 1982).

In very hot weather, the body temperature of a giraffe rises slowly during the heat of the day and falls again at night when the air is cool. A chart of the daily fluctuations of rectal temperatures shows midday body temperature of perhaps 40°C and midnight body temperatures down to 34°C (Langman, 2012).

However, as Vaughan Langman *et al.* (1982) note, raising one's body temperature only works if one is a large animal. Small animals in the desert in summer would overheat immediately if they stayed in the sun, so to survive they must always seek shade, usually underground. What about a giraffe calf, then? It isn't very large, and indeed it too has a problem with hot weather. This seems to be why it spends much of its first months, or even its first year, staying hiding in the shade during the day while its mother might range as far as 15 miles away to find forage. She will visit it several times daily when it is young so it can nurse, but less often as it grows.

Sweating?

There was controversy over whether giraffe do or do not sweat. Giraffe could keep cool in part by sweating, but by doing so they would lose valuable water from their skin. Mitchell and Skinner (2004) maintain that sweat glands are scattered through the skin of giraffe, but are more numerous, more dilated and more active under the patches (spots), especially on the trunk where there are also more blood vessels, and that they secrete sweat into hair follicles. The patches serve therefore not only to camouflage a giraffe but as a thermal window that can be activated in very hot weather (Mitchell and Skinner, 1993); infrared thermography shows that the spots radiate more heat than does the white background even long after the sun has set (Hilsberg-Merz, 2008).

Langman (2012) agrees there are sweat glands in the skin of giraffe, but has found categorically that they are remnants of the ancestral giraffids from which giraffe evolved, and that they are not functional. He also notes that:

- giraffe have evolved formidable other adaptations for saving body water;
- sweating in large species is difficult because of the body mass to surface ratio; most megaherbivores don't sweat;

because of his discovery of the increase of body temperature of camels during hot weather to conserve water. Schmidt-Nielsen had also turned his attention to giraffe, which had evolved the same mechanism. Hilde and I had just returned from Mauritania where we had been studying, under extreme heat conditions, respectively, camel feeding/metabolism and the filming of camels walking, pacing and galloping (Dagg, 1978). We were writing together a book to be called *The Camel: Its Evolution, Ecology, Behavior, and Relationship to Man*, which was published in 1981.

- should a giraffe become sweaty by running in cool weather it might become too cold;
- much of the skin is so thick it acts in part as a thermal barrier;
- they drink little, indicating they are a water-conserving species; and
- zoo animals with access to water still show daily bouts of heat storage.

Giraffe are obviously superbly designed to live in hot climates. However, as we shall see in Chapter 10, they are vulnerable to cold weather, sometimes dying in temperatures well above freezing.

9 Pregnancy, growth, reproduction and aging

Pregnancy

Giraffe mothers get a bad rap. In zoos they sometimes turn against their new baby, step on it by mistake or refuse to nurse it. In the wild they have been accused of leaving their infants alone and vulnerable in the bush. However, as we see in this chapter, giraffe cows are admirable, hard-working mothers even if their behaviour is sometimes suspect when they are forced to live in captivity. Most are already pregnant again while tending to and nursing their current youngster, so there is little respite during their altruistic motherhoods. They give birth and nurse these young until they die. And when an infant itself perishes, its mother may be deeply upset, as we shall see. An adult cow giraffe comes into oestrus for the first time when she is about 3 years and 9 months old, and she continues cycling every 2 weeks until she becomes pregnant (Berry and Bercovitch, 2013). A female mates with a male when she is in oestrus but at no other time. In Zambia, the average age of first parturition for Thornicroft's giraffe was 6.4 years ($n = 6$). The oldest cow known to give birth was about 24 years of age; she disappeared a year later (Bercovitch and Berry, 2009b).

The average gestation period for giraffe is about 446–457 days (del Castillo *et al.*, 2005), although Rothschild's pregnancies in captivity lasted about 470 days ($n = 6$) (Lueders *et al.*, 2009b). About 3 weeks after giving birth, a giraffe comes into oestrus again while she is still nursing her calf, so theoretically she could bear and raise 10 young in her lifetime if one were born every 18 months. Cows do continue to reproduce for the rest of their lives, with an average interbirth interval of 22.6 months for Thornicroft's and Masai giraffe (Leuthold and Leuthold, 1978; Bercovitch and Berry, 2009b). After giving birth, three females became pregnant in 19, 23 and 27 days, but of course continued suckling their recently born calf (Hall-Martin and Skinner, 1978). Twins are rare, and both seldom survive. Twins that have thrived were born in 2007 at the Bioparc Zoo Doué in France (Gay, 2009); one nursed from its mother while the other was bottle-fed.

A few females are infertile. This was true of Cheeky, an 11-year-old cow at the Taronga Zoo who never gave birth (Dagg, 1970b). She, unlike any other cows there, was a target for the three bulls who often hit her hard with their horns. (She hit them back sometimes as well as she could, but she was much less strong.) Two females in Zambia who were often observed over 11.5 and 20.5 years, respectively, but never with calves, seemed also to suffer from reproductive infertility (Bercovitch and Berry, 2009b). In a zoo wanting a calf yet lacking a male, females can be inseminated artificially (Foxworth, 1997).

Births

There is voluminous documentation on actual births of giraffe in zoos (Dagg and Foster, 1982). A pregnant giraffe begins to appear so several months before her calf is born. At about 14 months gestation, her udder swells and may leak milk. Sometimes the shape of the mother's abdomen shifts as the foetus moves about within her body (Kristal and Noonan, 1979). About a week before the birth, her sacral muscles above the tail relax, and sometimes occasional muscle contractions of the uterus are visible. After closely observing the behaviour of a giraffe for 3 weeks before she gave birth, Mark Kristal and Michael Noonan (1979) believe there are few, if any, behavioural or physical changes that can predict the actual day of delivery.

When she is about to give birth, a female is restless, paces about and eats little although she may drink at length. She may lie down, but stands again when the birth is imminent. After the foetal membranes burst, the forelegs of the calf appear and then its head resting between or beside them. During labour contractions the mother often stands with her hind legs apart and her head straining forward. Eventually the calf falls 2 m or so to the ground; it is only after the umbilical cord is broken at this point that the calf begins to breathe.

Labour may last as little as 20 min or as much as 3.5 or even 15 h (Williams et al., 2007). If the mother is bothered by observers, it may prolong the time by many hours, in which case keepers may place a rope around the calf and pull it out of the mother or do a caesarean section as described later. About 30 cm of umbilical cord remains hanging from the calf for several months. The mother licks her calf free of the foetal membranes, but she may or may not eat the afterbirth which is expelled any time up to 9 h later. The Toledo Zoo which has experienced various birthing problems among their giraffe is willing to share their expertise with other zoos (Miller, 2012).

The new calf in a zoo is about 1.8 m tall (range 1.3–2.1 m) and averages about 55 kg in weight (see data in Dagg and Foster, 1982, Chapter 10), apparently less than a calf in the wild. In Mpumalanga, South Africa, four neonate wild Cape giraffe weighed an average of 101.2 kg (Skinner and Hall-Martin, 1975).

Normally a neonate struggles to its feet when about an hour old, searches for one of its mother's four teats and begins to suck. She may nudge it to help it find the way. During the next few days the mother and young often sniff and touch each other as they bond. If a calf is born prematurely, the mother's maternal instincts may not yet have come into play and she may injure her young by accident, as if not realizing what this creature is. Primapara or disturbed mothers may also have trouble relating to their full-term youngster to the point of injuring them or allowing them to die. But even if they are sickly when born, calves can sometimes be saved, including one male who could not suckle, but was fitted with a nasal feeding tube that saved his life (Rouffignac, 1997). Another newborn also was at first unable to nurse naturally, but was able to feed from a bottle. By holding the bottle near his mother's teats, the staff at the Cheyenne

Mountain Zoo gradually switched his suckling from the bottle to the teats (Dillon and Bredahl, 2010). After a week of supplementary feedings, the bottles were phased out and the youngster suckled entirely from his mother from then on. A calf has been successfully hand-reared at Miami Metro Zoo (Webb, 2008), as have calves in Spain and Germany (Casares *et al.*, 2012).

Dagg and Foster (1982) list early behaviours and characteristics of neonates as documented by scores of zoos. Usually the calves are equally likely to be male or female, although Bourlière (1961) notes that of 117 giraffe born in captivity, 61.5% were males. In some species a mother invests more energy in one sex than the other as an evolutionary strategy, but this may not be true for giraffe. In general they have an unbiased sex ratio and similar interbirth intervals following the birth of male or female young (Bercovitch *et al.*, 2004). However, in a stressed population, more females may be born than males (see Chapter 4).

Timing of births

In the wild, birth dates of giraffe calves may be spread throughout the year. This was true in Tsavo East National Park, Kenya, where the births of 31 giraffe occurred in every month, just as they did in nearby Nairobi National Park (Leuthold and Leuthold, 1975), in Arusha National Park (Pratt and Anderson, 1982) and in Zambia (Berry and Bercovitch, 2012).

However, as A. J. Hall-Martin, J. D. Skinner and J. M. van Dyk (1975) noted, these giraffe live near the equator. In South Africa where seasonal changes are more marked, they found that although a few giraffe were born year-round, 60% of conception dates of 123 giraffe calves and 20 foetuses occurred during the four months of December to March. At that season rainfall was highest and food plants most nutritious and abundant, so there did seem to be a bias toward optimum environmental conditions. This bias favours the optimal condition of female giraffe when they become pregnant, whereas in other African ungulates it is the newborn itself that is born at the best time of year for them to prosper. (However, in this area, other giraffe births are reported to have peaked in the periods September to October and February to April; Owen-Smith, 2008.)

Giraffe born in zoos far from the equator also seem to favour spring conceptions. Captive giraffe may give birth in any month, but 86 giraffe born in zoos in Sydney and London tended to reproduce more in the spring than at other times (Strahan *et al.*, 1973) – in London over half occurred between March and June while in Sydney over half took place between July and November.

A. J. Hall-Martin and his colleagues (1975) also checked out the reproductive organs of males to see if they had seasonal changes, but they could find none. Seminiferous tubule diameters fluctuated with age, as did testes mass and epididymides mass, but not season. Testicular testosterone concentration varied from trace amounts to 10.1 μg/g of testicular tissue.

Figure 9.1   Female Angolan giraffe in probably a crèche situation as the calves are of different ages rather than twins. Photo by Julian Fennessy.

Growth

First weeks

To determine a model of mother–calf behaviour during the first 3 weeks of a neonate's life, the late Thomas Greene, S. P. Manne and L. M. Reiter (2006) collected extensive observational data on three female giraffe and their six calves born in the Brookfield Zoo, Illinois over a 12-year period. They established a baseline activity profile of cow–calf behaviour so that other researchers and keepers could judge if their neonates were developing normally. In zoos, for example, they could determine if a female was rearing her calf adequately or if it should be removed from her care to be hand-fed by keepers.

After the birth, calves and their mothers were kept in a separate stall for a week before joining the other giraffe. Suckling increased to 5% of the neonate's activity budget by week three, while 'grooming and investigation' (= licking itself and sniffing around the enclosure) increased to 25% by the third week. By contrast, the categories of standing still and lying down at rest fell from 44% to 34% and from 31% to 26%, respectively, during the three weeks.

When a 3-month-old calf at the Buffalo Zoo was separated briefly from its mother for an examination in a closed room, it gave a loud bellow (Kristal and Noonan, 1979a). Its mother immediately began kicking at the door with her front feet and bucking at it with her hind legs, although she did not vocalize. She struck out at the door again and again for 5 min, after which she paced back and forth in agitation. When the calf was released about 20 min later, she immediately began to nuzzle and lick it.

At the St Louis Zoological Park in Missouri, the udder of a reticulated female who had given birth 3 days earlier became so engorged and painful that she allowed her baby to nurse only briefly (Fischer *et al.*, 1997). On day 10, because the calf's weight was falling, the female was tranquilized lightly enough with xylazine that she allowed her calf to nurse. This procedure was repeated three times at intervals of 2 days, and then every day for 28 days, with the optimal dose of the drug 125 mg. By day 50, tranquilization was ended because the calf was now nursing normally without hurting his mother.

At another zoo where a female rejected her newborn calf, the staff was reluctant to raise it by hand because this can lead to gastrointestinal tract problems, infectious diseases and growth abnormalities (Borkowski *et al.*, 2010). The youngster refused to bottle-feed, but luckily there was another new mother giraffe at the zoo who allowed this calf to suckle along with her own young. This worked well. Both calves were successfully raised by the dam, remained healthy and eventually were weaned at about 6 months of age without human interference. Both calves not only got along well with the mother, but they also socially bonded with each other. If necessary, a newborn calf can be reared by zoo staff members, but this is 'a challenging, time-consuming and labour-intensive process' (Casares *et al.*, 2012).

Michelle Miller and her team (1999) have researched the passive transfer (or failure thereof) of immunoglobulins in neonatal giraffe.

Early months

In the wild, females tend to give birth alone but a few days later the two join up with other giraffe. Neonates sleep well at night, walk near their mothers and sometimes gallop even faster than she does. They can soon look after themselves, living for a few months or so on milk before gradually adding leaves to their diet. In the Serengeti, David Pratt and Virginia Anderson (1979) timed 415 calves as they suckled (for a range of 4–226 s, a mean of 66 s and a median of 55 s), most often between the hours of 18:00 and 19:00. The nursing regime began with sucking 1.1 bouts/hour in the first week, 0.65 in the second week, 0.30 for the second fortnight through the fourth month and about 0.2 bouts/hour from then on. Suckling tapers off in the second year, certainly well before a new calf is born. However, at the Kigio Wildlife Conservancy, a male Masai giraffe about 3 years old was observed suckling from a female about 12 months pregnant. She allowed him to nurse for about a minute before gently pushing him away (Anon., 2012g).

Nursing is initiated either by the calf or its mother who may at intervals approach and stare at her youngster. He or she seldom refuses this invitation. The mother is more likely than the calf to terminate a session by moving her hind leg or walking away. Other giraffe may try to nurse from the mother too, but she does not allow this.

Various analyses of giraffe colostrum and milk are available (Stephan, 1925; Aschaffenburg et al., 1962; Ben Shaul, 1962; Skinner, 1966; Glass et al., 1969; Hall-Martin et al., 1977b; Dagg and Foster, 1982). Colostrum is very high in fat, nonfat solids and protein. This first milk after birth has more total solids, fat, ash and protein than later milk, but less lactose. Its importance to neonates is vital. One calf born at a wild animal park never nursed from her mother but was removed when a day old and fed whole homogenized cow's milk and vitamins. She never had colostrum (Jacobson et al., 1986). Three weeks after birth, she had to be treated for infectious polyarthritis and polyosteomyelitis in her joints. Three painful months later she was sent to an animal hospital where she died. Colostrum, which confers on a neonate immunity from some diseases, should always be given to newborn calves, even if it comes from another species such as a domestic cow.

Compared to bovine milk, giraffe milk has more fat, total protein and ash content, but less lactose. Casein, lactalbumin and lactoglobulin fractions of the milk proteins are similar in the two species. Lactating females produced between 2.5 and 10 litres of milk a day. In Niger, evidence indicates that lactating females avoid eating leaves with a high tannin content that might harm their calves (Caister et al., 2003).

A neonate is a young giraffe up to a month old. A calf (between ages one and two) regularly accompanies its mother, while a juvenile or subadult is a youngster who is on its own. At 5 years of age, a cow or bull is considered to be an adult (Pratt and Anderson, 1982). A female can give birth at age five, but a male does not become sexually mature or stop growing for several more years. Growth curves have been worked out by S. M. Hirst (1975) for males (78-kg neonates to 1200 kg) and females (70-kg neonates to 800 kg).

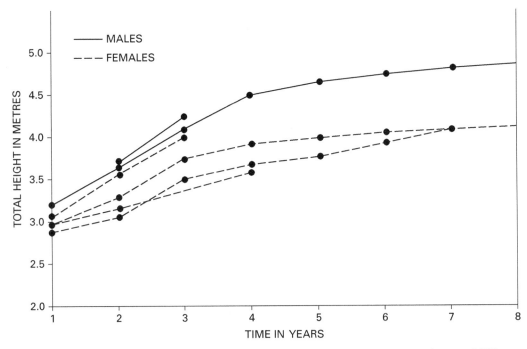

Figure 9.2 Yearly growth rates of captive giraffe from ages one to eight. From Dagg and Foster (1982).

Chromosomes

The karyotypes of two giraffe from Kruger National Park have been reported by Wallace and Fairall (1965). The modal number of chromosomes present in 31 chromosome counts was 30 chromosomes per cell. The karyotype of the giraffe is similar to that of bovids, especially the sitatunga (Koulischer *et al.*, 1971). The Y chromosome was metacentric rather than acrocentric.

David Brown and his colleagues (2007) have analysed mitochondrial DNA sequences and nuclear microsatellite loci for giraffe in most areas of Africa as part of their research into the speciation and subspeciation of this species.

Female anatomy

The right oval-shaped ovary of adult females is about 24 mm by 35mm, with the left ovary slightly larger (Lueders *et al.*, 2009a). Follicles of varying sizes are present in immature, cycling and pregnant animals, but a Graafian follicle and its ovulation occurs only in cycling females. The histology, ultrastructure and progestins of ovaries of foetal and neonate females are described, along with their position *in situ*, by Kellas *et al.* (1958), Kayanja and Blankenship (1973) and Gombe and Kayanja (1974), and the uterus by Hradecký (1983a). Implantation occurs on the same side as the corpus luteum of pregnancy and development can proceed for at least 30 days before the embryo becomes attached to the uterine wall (Hall-Martin and Skinner, 1978). The placenta, which weighs about 8 kg, has many cotyledons, the largest measuring 23 × 5 cm and the average 6.4 cm in diameter (Wilson, 1969); it has also been described by Deka *et al.* (1980) and Hradecký (1983b). The umbilical is about 1 m long which is why it breaks when the calf falls to the ground at birth. A. Murai and colleagues (2007) report on a teratoma of the umbilical cord.

Male anatomy

The reproductive organs of the male giraffe resemble those of other ruminants in anatomy and histology. The penis, which does not contain a bone, is enclosed in a sheath from which it emerges when it is in erection. Its anatomy, the testes and related glands are described by Retterer and Neuville (1914), Neuville (1935) and Glover (1973). Spermatozoans of an adult have an average head length of 5.77 μm and an average total length of 47.77 μm (Velhankar *et al.*, 1973).

After spermatogenesis occurs in a male at age three or four, there is a rapid increase in size of the testicles. In the foetal testis the dominant hormone is androstenedione while in the adult testis it is testosterone (Hall-Martin *et al.*, 1978). The testosterone concentration in the testes of 19 males killed over a period of one year did not vary with the season, but did increase with increasing age (Hall-Martin *et al.*, 1975).

Reproductive physiology

Zoos are a good place to study medical issues of giraffe; the age, history and often race of the animals are known, they are used to the presence of people and they can be trained to accept some restraint in return for favoured foods. Imke Lueders and six colleagues (2009a) worked with one male and six female Rothschild's giraffe at the African Lion Safari near Cambridge, Ontario, Canada, and a male from the Melbourne Zoo in Australia. Over the course of the research, females would be classified variously as immature, cycling (nulliparous or parous), pregnant or postpartum.

The females had daily access in warm weather to a 70-acre outdoor display area. For part of their training process, as they filed from their night quarters into the outdoors each morning, they passed through a narrow hallway-cum-restraining area. There, after considerable habituation, a cow could be stopped by sliding doors and fed food while transrectal manipulation and faecal sample collection were carried out. (Earlier, urinary samples would have been used for pregnancy tests but these are tricky to collect; Loskutoff *et al.*, 1986.) Also, an ultrasound probe was inserted into her vagina by a person standing on a ladder so all parts of the inner reproductive organs could be scanned from various angles. This process was done one to four times per week to ensure the collection of ample data. The reproductive organs of the male were also scanned by a lubricated probe passed into the rectum.

The results of these examinations are given in detail in the report, with brief generalizations as follows:

(a) ultrasound can accurately depict internal genital structures of female and male giraffe and thus is an invaluable tool for reproductive tract assessment, breeding management and health control;

(b) stages of follicles, pseudo-corpus lutea and corpus luteum are described and can be correlated with faecal analyses; and

(c) with ultrasound, pregnancy could be detected as early as 28 days after mating.

In a follow-up paper, Imke Lueders and colleagues (2009b) examined the ovarian activity of these same females when they were cycling and when they were pregnant, in particular correlating faecal progestins and oestradiol during the short oestrous cycle (15 days) and early pregnancy. Variations reflect the short inter-oestrus periods in females which enable them to optimize the chance of mating with roaming solitary males. The authors document the precise time of ovulation which is useful to know for zoos concerned with reproduction (although findings may not be applicable to all races of giraffe). Early ultrasonographic monitoring of ovarian activity and hormonal manipulation is reported by Gilbert *et al.* (1988), an evaluation of progesterone levels in the faeces of captive reticulated giraffe by Dumonceaux *et al.* (2006) and faecal progestagen and oestrone changes during pregnancy by Isobe *et al.* (2007).

The behaviour of seven female giraffe living at the San Diego Wildlife Park was correlated with their ovarian activity (del Castillo *et al.*, 2005). The easiest part of the study was collecting faecal pellets four or five times a week from each giraffe; these were analysed to determine if their chemical composition reflected the ovarian condition of the producer. It did, to the delight of the researchers, because this meant they could follow the pregnancy and cycling of an individual without upsetting her and thus affecting her behaviour. The research hoped to understand the metabolic cost of being a mother; most large mammals are seasonal breeders, so it is impossible to separate changes caused by reproductive processes from seasonal changes in forage quality and availability. By contrast, the giraffe reproduces throughout the year, so her reproductive state can be uncoupled from seasonal changes that affect diet, habitat and activity. However, she is also unusual because she is lactating through most of her pregnancy.

Using the information provided by the faeces, the giraffe were categorized as either pregnant, acyclic or cycling, with each period clearly defined. There were other findings for the four females who bore surviving young:

- the mean number of days was calculated for each of the three reproductive phases;
- the average length of their gestation periods was 448 days;
- the acyclic period was significantly shorter for females whose calves died within one week of birth;
- the period from the resumption of cycling until the next pregnancy varied from 6 to 236 days;
- faecal pregnane concentrations increased steadily as pregnancy progressed, then plummeted after parturition;

- using a single faecal sample, pregnancy could be detected two weeks after mating;
- the four females with surviving young began cycling again from 39 to 142 days after giving birth; and
- unlike giraffe in the wild, the San Diego giraffe did not eat more nor different foods when pregnant. However, they did interact less with other giraffe and spent less time moving about.

Female giraffe have evolved a reproductive strategy which enables them to channel their resources into two offspring by differential timing should they become pregnant shortly after giving birth. While a neonate is consuming rich milk during its early months, her next young is developing slowly, the foetal mass weighing only 2 kg 150 days after conception. The weight of the embryo triples during the third trimester, however, indicating that after lactation decreases and ends, the mother's resources become devoted primarily to gestation (del Castillo *et al.*, 2005).

Also at the San Diego Wildlife Park, research has been dedicated to discovering a better reversible contraceptive for female giraffe (Patton *et al.*, 2006). Zoos might not have room for more calves at any one time, but later more young might be desirable if giraffe had died or if a new larger facility had been built. After implanting two reticulated females with the long-term GnRH agonist deslorelin (DESL) (which had proven effective for cattle and other ungulates), the researchers monitored their endocrine status using steroid analyses of their faeces; the group had earlier validated an assay for faecal pregnanes as indicated above, and this research validated an assay for faecal oestrogens too. One animal was a 5-year-old primiparous female and the other a 15-year-old multiparous ($n = 8$) cow. Both weighed about 600 kg.

The effects of the DESL differed in the two giraffe. The older animal exhibited sexual behaviour at similar rates to cycling females but did not become pregnant. Nor did the younger one who had little sexual activity. The researchers preferred DESL to the common progesterone-derivative medroxyprogesterone acetate (MPA) because it was non-steroidal and did not have to be injected every 6 weeks, which stressed the animal. After the experiment was complete, the younger giraffe continued on contraceptives and has not become pregnant, but the older one who was not given further implants produced a baby in 2008 (Meredith Bashaw, pers. comm.). DESL therefore has been shown to be a reversible contraceptive.

Until recently, a caesarean section had never been successfully performed on a pregnant giraffe. This changed when D. C. Williams, P. J. Murison and C. L. Hill (2007) were able to save a female giraffe who had been in labour for 28 h. She had a dead calf stuck in her vagina (dystocia). Zoo personnel were unable to dislodge the carcass despite using copious lubrication and traction, so they were forced to anaesthetize the female, keeping her head above the level of the rumen, and operate on her right abdominal flank. The hair was clipped from the rib cage to the upper thigh and from the lumbo-sacral processes to the ventral abdomen. This large area was then scrubbed with antiseptics before a 50-cm incision was made to remove the body and the placenta. Four personnel oversaw the operation which is fully described in their report. The female recovered normally, with the stitches removed after 3 weeks. She had possible matings 47 and 131 days after her operation, and a definite mating after 154 days. Other cases of dystocia have been described by Citino *et al.* (1984), Grunert *et al.* (1988), Juan-Salles *et al.* (2008) and Kaitho *et al.* (2011).

A. J. Hall-Martin and I. W. Rowlands (1980) collected the ovaries from 34 giraffe (and 4 foetuses) ranging in age from 2 weeks to 20 years which had been culled in South Africa. They describe in detail the anatomy of these organs.

Male sexual activity

Male giraffe become reproductively mature when 7 or 8 years of age (Bercovitch and Berry, 2012). The aim of adult bulls is to mate with females in oestrus, so they spend their lives largely alone, searching out groups of cows, hoping that one or more will be willing to mate that day. Ideally a male will spend minimal time with females not in oestrus and walking long distances, and maximal time with females who are in oestrus. For giraffe this roaming strategy, common in other non-seasonal breeding mammals as well, has three major costs:

(a) travelling great distances requires high metabolic expenditures;
(b) such travelling alone increases the risk that a male may be killed by lions; and
(c) by staying close and guarding one female in oestrus, a male misses the opportunity to mate with other females.

Dominant bulls (the A group of Pratt and Anderson, 1985) are not known to fight with each other over females in oestrus with whom to mate, so it seems that their reproductive strategy of roaming is that of endurance rivalry (Bercovitch *et al.*, 2006).

At the San Diego Wild Animal Park in a 36.5-ha enclosure, Fred Bercovitch and colleagues (2004, 2006) were able to observe mating behaviour among six adult Rothschild's females and two adult males over a 2-year period using urine samples for endocrine analysis of the giraffe ovarian cycle; urine revealed a cycle of 15.5 days, while faecal samples, the ones used in this study, indicated one of 14.7 days.

If a male suspected a female was in oestrus, he followed her and nuzzled under her tail to urge her to urinate. If she did so, he caught some of the urine in his mouth to flehmen or urine-test it. (Flehmen occurs in many mammals when the male inhales with his mouth open and upper lip curled back so that a scent can reach the vomeronasal organ which then verifies if the female is in oestrus.) The females were choosy though; of all the Masai bulls' attempts to test their urine in the wild, the dominant A bulls were most successful (61% of the time) and the least dominant C bulls least successful (34% of the time) (Pratt and Anderson, 1985). The C bulls' low success rate was partly because they were inept at nuzzling a cow's rump and partly because she chose not to urinate for them, especially if there were an A bull nearby whom she might fancy. All the bulls were much keener to test young cows than juveniles or older animals; one young female's urine was tested 17 times in 2 h, but such persistence was never geared toward females of other ages. Despite the recalcitrance of many females, urine testing was common, viewed 304 times over several years of giraffe observation, although it never involved a juvenile male.

Figure 9.3   Three Masai males follow a female who is likely in oestrus. Photo by Karl Ammann.

Sometimes a male tasted urine from a female who happened to be urinating anyway, or several males tested urine at the same time (Pratt and Anderson, 1985).

If a female is in oestrus, a male stands close behind and in alignment with her, sometimes with an erection, his pink penis visible beyond the penis sheath. He periodically taps the female's hind leg with a foreleg. Sometimes she moves away, in which case he follows closely behind her, or sometimes she stands still. If the latter, he mounts her, sliding his front legs up on either side of her torso. She moves forward at this, so emission may or may not be completed. If it is successful, ejaculation occurs immediately and lasts about 2 s; the semen volume is about 4–6 ml. The shortest time between mounting contemplation and first mating attempt was 20 min, and no male ever gave up the possibility of mating after becoming so involved. The least time recorded between one ejaculation of a male and the next was 1.5 h. There is no evidence that a female in oestrus refused to mate with a dominant male; to do so might mean that she continues to cycle rather than become pregnant. Certainly she could easily do this if she refused to stand still for the male to mount her. However, if she does play hard to get by continually walking away, it is assumed that she has not yet reached full oestrus status. In the Willem Pretorius Game Reserve in South Africa, a dominant male followed a female for at least 3 h and made several unsuccessful mounting attempts before he was successful in coitus (Kok, 1982).

Figure 9.4   Male reticulated giraffe has collected urine from a urinating female and flehmens to determine if she is in oestrus and therefore willing to mate. Photo by Zoe Muller.

After mating, the male usually wandered away, but not if there were a second male waiting his opportunity. Then the first male usually remained guard by the female so the second male could not mate with her in turn. The males did not fight each other over access to females in oestrus. In general, males were most likely at any one time to be near females who were pregnant or cycling. The authors found a four-day fertile window for females who were cycling; the males spent significantly more time near females in this phase and in their periovulatory period. By contrast, females directed affiliative behaviour toward males whether or not they were in their fertile window.

In the wild, females regularly have several cycles after parturition before they become pregnant again. Bercovitch and colleagues (2006) postulate possible reasons for this, such as the loss of a baby by the female, her state of health or her age, but probably it is a matter of meeting up with a male at the right time. Issues keeping males and females apart include frequent sexual segregation, changing female subgroups, large home ranges, dispersed food sources, lengthy gestations and the narrow fertile window of females during each cycle. When a dominant male does join a group of giraffe, he examines the anogenital region of all the adult females in it. The frequency of sexual investigations is nearly three times as high during the fertile window period of a female compared to the rest of her ovarian cycle, and eight times as high as when the female is pregnant. The authors conclude their study as follows: 'The precise cues emitted by females, the perceptual mechanisms adopted by males, and the possible links between male ability to detect cues and their internal hormonal milieu remain to be determined.'

In general, when an adult male joined a mixed group of giraffe he was somewhat interested in the calves. He typically approached each small youngster in turn and bent down to nose it, usually on the head (Pratt and Anderson, 1985). In response, the calf might touch the bull's head in return, walk or run away, or, if very young, hunt for a teat. Then the male turned to his main pursuits of testing the reproductive condition of each adult female and of browsing.

Castration

Zoos may castrate males if they are not needed for breeding, come from a hybrid lineage (and therefore will not be wanted for studbook genetics) and are creating stress in their group by their fighting. This latter happenstance involving two intact males among other neutered males occurred in the Port Lympne Wild Animal Park, UK (Chamberlain, 2012), so they were castrated following careful preparation:

- readying a roomy area with padding, cushioned floor and no objects such as feeders to prevent both pre- and post-operation injury;
- discussion of safety concerns including exits for people from the area in an emergency;
- training of each animal so that he is used to the set-up prepared for him; and
- arrangements for post care processes for each male.

Large mammals sometimes die following surgical castration, but this operation was successfully performed on three subadult males weighing 555–711 kg (Borkowski *et al.*, 2009). Surgery involved an emasculator plus ligation. One male developed scrotal dermatitis, but there were no other complications. Thiafentanil, medetomidine and ketamine plus local lidocaine worked satisfactorily as anaesthetics.

Frantic mothers

Males do not know which young they have fathered, but females who do know their own progeny fight off lions or other predators to keep them safe. If their calf dies, they can be distraught. Zoe Muller (2010a), head of the Rothschild's Giraffe Project in Kenya, oversees the daily activities of about 65 giraffe in Soysambu Conservancy. She worried about one month-old calf who, because of a badly deformed hind leg, had great difficulty walking. It spent most of the time standing still with its mother, F008, never more than 20 m away.

One morning on her daily tour of the property, Muller came upon 17 females running excitedly around near a bush in vigilant behaviour and as if in distress. On driving closer

Figure 9.5   Rothschild's mother and a friend investigate the body of her dead calf hidden in the long grass. Photo by Zoe Muller.

to the centre of their attention, she saw F008's calf on the ground who had just died of natural causes. Muller retreated in her vehicle to a distance to watch the action.

For the next 3 h that morning the females milled about, showing extreme interest in the body by continually approaching and retreating from it. When Muller returned in the afternoon, there were 23 females and 4 juveniles present, all restless and walking about near the carcass. F008, other females and juveniles were approaching the body at various times to nudge it with their muzzles.

The next morning F008 and six other females were near the carcass, circling around as if protecting it. That afternoon 11 females (including F008) and 4 large males were present, although the males kept far back from the carcass. Their interest was obviously in females who might be in oestrus, and in browsing. In the evening F008 and 2 females were positioned within 10 m of the carcass, all in vigilant behaviour and none foraging or ruminating – apparently still being guardians.

The following day, only F008 was there standing under a tree 50 m from where the body had been. She seemed to be keeping watch over the carcass, now half-eaten, which lay at her feet. F008 remained alone on guard for the rest of the day. The next day she was still in the area, but the remains of her youngster were gone.

A drama similar to that observed by Zoe Muller has been documented by Megan Strauss in Serengeti National Park, although this calf was a year old (Strauss and Muller,

Figure 9.6   Zoe Muller with Rothschild's giraffe. Photo by Sabine Bernert.

2013). The Masai mother and two females watched lions eating the carcass they had newly killed from over 200 m away. These three were present again the next day, although closer because the lions had left the bones and skin for hyenas. The mother was again near the kill site on day 3, but on day 4 she was several kilometres away with three new herd members. However, she returned to the area on day 11.

A maternal instinct is obviously present at the moment of birth itself, because a Thornicroft's giraffe who had yet to expel the placenta exhibited a close tie with her newborn calf which was apparently born dead (Bercovitch, 2013). Females are almost never seen alone, but she stayed near the carcass by herself for 2 h, splaying her forelegs to lean down and sniff the body while the rest of her herd wandered on, leaving her behind.

Another distraught mother resorted to cannibalism. O. Kok (1982) was aware of a Cape female giraffe who had given birth about a week earlier. After the baby died, she was observed walking back and forth near the corpse, staring at it, and several times bending over it to smell or paw at it. Eventually she started tearing and chewing meat from the exposed ribs. She repeated this odd behaviour 2 weeks later, and 3 weeks after that. On each of these occasions she left her small group to return to the body, smell it, and then repeatedly pick it up and chew on it for several minutes, although by this time the remains were dried up and cracked.

A Namibian giraffe in Etosha National Park also showed interest in the area where a young adult female giraffe had been killed. Kerryn Carter (2011a) reports that a camera was set up near the carcass to monitor scavenger activity in connection with a long-running survey of anthrax virus. Three weeks later, after the body had been removed by scavengers, the camera filmed a group of passing giraffe stopping to examine the site where the remains had been. First, a male leaned over the site to sniff and scan the ground. Then five other giraffe also inspected the site. Had these giraffe seen their friend die? Did the smell of giraffe attract them? Could they tell if this smell belonged to an individual they knew? Or were they looking for bones to chew, as is common? There is no way of knowing from the limited camera tape. (Such cameras should play an important role in future wildlife research.)

There is ample evidence that elephants and some social higher primates grieve the death of one of their members (Dagg, 2009), but giraffe, documented here for five races, are the first ungulates seen to be equally concerned with death, and indeed perhaps more concerned because several remembered the spot their young had died for many days. It may be that such emotion is more common in mammals in the wild than we yet appreciate.[1]

Older males with darker spots

The age of a male giraffe can be judged to some extent by the colour of the spots on his torso. Phil Berry and Fred Bercovitch (2012) have collected information about coat colour in 36 Thornicroft's bulls over a period of 33 years. All spotting of giraffe darkens with age, but it is notably so in males. Their body spots change gradually from a brownish to black colour, taking 1 or 2 years to complete the transition. The spots darken first in the centre and then gradually outward; when they are completely black, the male is considered to be on average an adult of about 9.42 years (range 8.50–10.57, $n = 4$).

---

[1] Birds also may respond to a dead conspecific. For example, a western scrub-jay (*Aphelocoma californica*) will give an alarm call at seeing a dead scrub-jay which will attract numerous other noisy scrub-jays to the area (Iglesias *et al.*, 2012). (By contrast, a scrub-jay skin mounted in an upright, life-like posture elicited aggressive responses, indicating that other birds thought it was alive, unlike the dead bird.) The loud noise made by the aggregation of birds perhaps deters further attack by the presumed predator that had killed the bird.

Longevity

Berry and Bercovitch (2012) worked out the average life span of a male giraffe in their Thornicroft population who had survived his first year to be from 14 to 16 years, with a maximum span of just over 20 years. This is less than that calculated for the Masai race at 25 years (Dagg and Foster, 1982). On average, a bull lived for 7.43 years after attaining maturity while the females lived somewhat longer, the oldest until about 24 years of age. The males therefore have on average about 6 years of sexual activity, while the females can produce young from age 6 until at least age 24. Zoo giraffe can live to be 20 or 25 years of age (although many die prematurely because of their captive conditions); one Rothschild's captive, Clara, has lived for at least 34 years at the Bronx Zoo (Powell, 2010).

Do males lose interest in sex with age? This seemed to be the case for the 24-year-old male Jan at Taronga Zoo (Dagg, 1970). Barbara Leuthold (1979) also thought several old Masai males in Tsavo East National Park were too old to mate. However, in Arusha and Tarangire National Parks in Tanzania, three males identified by their appearance as 'very old' were still in the game (Pratt and Anderson, 1985).

Mortality

For their study of Masai giraffe in Tsavo East National Park in Kenya, Barbara and Walter Leuthold (1978a) report on giraffe mortality for the animals known individually to them. For male and female adults and subadults, the annual mortality rate was about 10%. (This mortality rate is actually a rate of disappearance, because the giraffe presumed dead were not again encountered by the Leutholds, so the number is approximate.) Of 15 calves born, 5 survived to 1 year of age (33%) and 4 to 2 years (27%). For young older than 1 year, males died at a greater rate than females. As this Park has few roads, the Leutholds often did not see a female for months at a time, so one may have given birth and her baby died without them knowing about this. For Thornicroft's giraffe, Bercovitch and Berry (2009b) found that about 45% of young died in their first year. The list by Foster of the number of giraffe present at different ages in a population (Chapter 4) gives a good idea of the mortality rate of each age group, given that similar numbers of males and females are usually born.

10 Giraffe in zoos

Giraffe are good for zoos because they attract people intrigued by this amazing species. They are bad for giraffe, though, who are used to wide open spaces and a huge quantity of various leaves on which to munch. Fortunately, zoos have basic standards for animal welfare: animals must have freedom from disease, the possibility of reproducing and a long life (Bashaw *et al.*, 2001).[1]

Recently, the importance of psychological welfare has also been advocated which involves the reduction of negative stress, boredom and trauma for animals. Animals in the wild encounter new experiences every day; if possible, they should have similar stimulation in captivity. These standards apply to all species, including giraffe.

To some extent, the health of a giraffe can be estimated from what it looks like. Members of Disney's Animal Kingdom asked North American facilities holding giraffe to send them digital photos of their animals so they could compare and score them as to body type and conditioning (Christman, 2008). Seventy institutions responded. Seven knowledgeable experts then rated the 100 most clear images on a scale of one to five, one being giraffe in the poorest (emaciated) condition and five the fattest/obese animals. This comparison made it possible for keepers to decide if their own giraffe needed a change in diet or other conditions, which they were urged to undertake cautiously and under the care of a veterinarian. Most obviously, having a thin neck and bony prominences indicate a giraffe in poor condition.

Captive giraffe are unique in a number of ways which will be considered in this chapter.

- Giraffe in North American and European zoos die prematurely in alarmingly high numbers. Recently, 75% of these deaths in the United States have been caused by husbandry, nutrition or management decisions, indicating that many of them should have been prevented (Gage, 2012).
- Giraffe have special dietary needs; in the wild they eat 99% browse, but such a diet is impossible to sustain in zoos, especially in winter. No one knows exactly what their nutritional needs are, but they certainly must be fed a palatable diet that meets their energy requirements.

---

[1] A long life is not necessarily a good thing, however. A zoo should consider putting down an animal who is in pain or no longer sentient (Paul Rose, 2012, pers. comm.).

- Some giraffe sustain injuries in zoos or during transport that are difficult to treat because anaesthesia is potentially dangerous for giraffe and itself causes deaths.
- Training giraffe works well. A major husbandry issue is overgrown hooves which affect giraffe movements; these and other ailments can be best treated if the animal is trained to stay calm in restraint.
- Because giraffe are unable to keep warm in cold weather, they may die if their diet is poor and their winter quarters inadequately heated.

Racial purity

With the recent understanding that races of giraffe have significant genetic differences and may in fact be species rather than subspecies, zoos are working to decrease their number of hybrid animals. If young are to be born, it is important that both parents belong to the same race. This endeavour has been fairly successful in Europe, with hybrid individuals now down from over one-third to fewer than one-quarter of the total European giraffe population (Damen, 2008, 2010). The Angolan and Cape giraffe are now facing extinction in European zoos, but the Rothschild's, reticulated and Kordofan giraffe are doing well.

The races of giraffe in North American zoos are varied (Brenneman, 2004). They were thought to be mostly reticulated, Rothschild's, hybrids of these two or Masai giraffe, but in a genetic analysis of 125 individuals (about one-quarter of the North American population), only about half the giraffe nominally of the first two races had genotypes consistent with that race in the wild; they may have looked like one of these races, but in reality half of them did not belong to them. The problem arose with the early 'founder' animals shipped into North America. In theory they were said to belong to particular races, but sometimes became mixed up when they arrived listed as simply 'of African origin', or young were of 'parentage unknown' or keepers confused animals of different races travelling together. No one was much worried about subspeciation 80 years ago, although today the consensus is that captive hybrids should not be allowed to breed.

Dietary concerns

Ideally, giraffe in zoos should have the same diet as they have in the wild. It is what their bodies are used to, and why they evolved the way they have. In zoos, however, it is impossible to feed giraffe entirely on leaves, which are expensive. Their food rather was largely based on that given to domestic stock; it aimed to make cattle grow quickly, reproduce as early as feasible and produce a maximum of muscle mass (meat), which is not what is wanted for captive giraffe (Schmidt et al., 2009). This diet translated into an unsatisfactory combination of available hay (grass hay and/or alfalfa hay) and a concentrate ration providing supplemental energy and micronutrients. If some browse was offered it was of European or North American plant species, which are not the same as those in Africa. The various ailments listed below in the 'Food-related ailments' section are thought to be connected with diet, so the food consumption of giraffe is extremely important, and indeed has been extensively studied (see Hatt *et al.*, 1998).

How to compare diets? One way is to analyse the composition of the blood serum of giraffe from the wild and from zoos. If they contain the same, or almost the same, components, and in the same proportions, the zoos involved would be elated. However, this is unlikely even in the best zoos. For giraffe in the wild:

- their health and past activities can only be guessed at when their blood is collected; phosphorus quantity, for example, could be related to their unknown exercise levels (Dikeman *et al.*, 2009);
- their blood sera will vary depending on the season and the vegetation consumed;
- giraffe are sedated to obtain their blood, which process probably affects their sera; and
- there may be problems in keeping the sera at suitable temperatures, or performing delicate laboratory analyses.

Nevertheless, this challenging research on comparative diets was carried out by Debra Schmidt and several teams, first in a pilot study of wild giraffe (2007) and then in a comparative study (2009) in which blood was obtained from 20 healthy giraffe from American zoos and from 24 free-ranging giraffe in South Africa. The aim once they obtained the data from the two sets of sera was to focus on changing zoo diets in a direction that would make the sera more similar.

The components studied were concentrations of amino acids, fatty acids, lipoproteins, vitamins A and E and minerals. Of the 80 parameters measured, 60% were significantly different between the two sets of sera. Nine items were significantly different between subadults and adults, and only one, sodium concentration, was significantly different between males and females. The authors note that the amino acid profiles seemed to reflect the greater activity of free-ranging giraffe and the consequent increase in protein turnover. To improve the giraffe diet, they suggest an increase in omega-3 fatty acids in zoo diets and a careful examination of retinol metabolism (an earlier study had found differences with age and sex too, and also a small difference between the sera of wild giraffe physically restrained and those anaesthetized; Bush *et al.*, 1980).

The second comparative study also headed by Debra Schmidt (2011) found that pathological conditions of giraffe in zoos may be associated with an increase in glucose, alanine aminotransferase (ALT) activity and bilirubin values. If the elevated ALT activities and bilirubin concentrations are related to subacute rumenitis, as suspected, there should be a change of diet for giraffe and also for other specialized ruminants. Ray Ball (2010) notes 'A central hypothesis is that dietary induced rumenitis and resulting changes in physiology are central to the disease syndromes seen in captive giraffe.' These affect food intake and its digestibility, milk production, hoof health and overall animal health.

Browse versus hay

A need for such research indicates that diet is a huge problem for giraffe in zoos. In the wild these animals are browsers, but in zoos browse is hard to come by, expensive and labour-intensive for the staff providing it. Browsers such as moose, bongo and okapi are often called concentrate selectors, but giraffe could be classified as a facultative concentrate selector or an intermediate feeder in part because they do not seem to select for a diet high in energy or low in fibre (Baer *et al.*, 1985). These browsing species are considered to be early evolved ruminants who are less able to digest grasses than are grazers. Browsers utilize the cell contents of forage, whereas grazers also utilize cell wall constituents. However, the digestive system of giraffe does have similarities with those of grazers (Valdes and Schlegel, 2012).

Traditional diets for giraffe in North America consisted mainly of low-fibre pellets and alfalfa hay, a regime designed for domestic ruminants and for the many grazing species in zoos; they were high in soluble carbohydrates (sugar and starch) and low in total fibre (Valdes and Schlegel, 2012). However, such regimes including considerable amounts of grains or hay may not provide enough nutrients for the good health and long life of giraffe; certainly eating only lucerne hay diets is unlikely to meet their energy requirements (Hatt *et al.*, 2005).

Over the years, it has become increasingly clear that the more browse captive giraffe consume, the more likely they will be to thrive. An optimal diet would consist entirely of browse, which is too expensive. Leaves of trees and bushes not only provide the best nourishment but, because they take longer to eat, they exercise the mouth and tongue and help prevent stereotypical behaviours. (See Stereotypical behaviours.) Browse for giraffe (and a few other species) is so important that in some countries browse farms have sprung up to supply the demand. In the future, silage of willow and other plants may extend the browse possibilities for giraffe, especially in colder weather when they have higher energy requirements. Ideally for zoos, a nutritional analysis of browse from local species of trees and bushes should be compiled.

Protein

Pellet concentrates high in protein (18%) and low in fibre were once common for giraffe as a response to Serous Fat Atrophy (discussed below) (Valdes and Schlegel, 2012). In 1973, for the first time, it was suggested that a better diet for browsers such as giraffe would be one higher in fibre and lower in protein. Too much protein in the diet can cause chronic acidosis, overgrown hooves, obesity and possibly a shortened life span. It took over 30 years for this superior regime to be accepted at most zoos. The amino acid profiles of sera in wild giraffe indicate a higher protein turnover, but of course they are far more active in their daily lives (Schmidt *et al.*, 2009). A diet too high in fibre is bad (Clauss *et al.*, 2001) – giraffe have a fairly weak rumen compared to buffalo, for example, and they don't ruminate as much (Rose, pers. comm., August 2011). Despite many consumption and behavioural studies of captive giraffe, 'their nutrient requirements remain unknown' (Valdes and Schlegel, 2012, p. 613).

Diet recommendations

Paul Rose, Jurgen Hummel and Marcus Clauss (2006) who analysed the metabolizable energy (ME) in a variety of giraffe diets in 7 British zoos holding 33 giraffe determined that they were adequate, but not great. There was a correlation between the amount of concentrates offered/ingested and the ME intake, but not between ME and roughage intake. The proportion of roughage in the overall diet was highly variable, depending on the forage quality. The recommendations of the authors were:

- the upper limit for concentrate should be no more than 40% of the diet;
- good feeding supplements are vitamin E, selenium and linseed (for linolenic acid);
- protein intake in concentrates should be perhaps 12%, not 18% (although more lucerne was eaten when forage protein was high);
- lucerne is not ideal for giraffe, but the best compromise currently available. Even so, giraffe cannot maintain their energy demands on a diet of lucerne hay; and
- some zoos occasionally supply their giraffe with novel food items such as peanut butter, apples, popcorn, pumpkin paste or biscuits, but such treats are high in sugar and bad for giraffe.

More work needed to be done on diets! Two teams answered the call. That led by E. A. Koutsos (2011) comprised eight scientists from five different facilities who collaborated to decide upon an improved diet. In their experimentation, six giraffe (ages 2–12) who had been fed a typical hoofstock menu were switched to diets containing reduced starch, protein, calcium, phosphorus and increased omega-3 fatty acids. This diet, along with a 50 : 50 mix with alfalfa and grass hay, was fed to them for 4 years. With their new diet the giraffe had serum with less saturated fatty acids and increases in omega-6 and omega-3 fatty acids; their blood nutrient profiles were more similar to those of wild giraffe than they had been on the hoofstock regime, so this was an improvement. The authors suggest that additional amounts of omega-3 fatty acids might be useful along with perhaps a further reduction in dietary starch and protein levels. Wild giraffe have nine times the absorptive capacity for volatile fatty acids (VFAs) as have some captive giraffe. VFAs are essential for ruminants because they provide up to 90% of available energy and are directly absorbed through the rumen mucosa.

Eduardo Valdes and Michael Schlegel (2012) also sought a better diet. After evaluating recommendations from four earlier feeding studies, they advocate diets for six groups of giraffe: males and females less than a year old, 13–24 months and over 24 months of age. For each category they include the average body weight and the ME needed each day, an invaluable resource.

Water

Fresh clean water should be available to captive giraffe at all times. In the wild giraffe may go for long periods without drinking if they have access to leaves with high water content. However, if water is available they often drink. In zoos where browse is limited, water is essential.

Food-related ailments

Like other ruminants in zoos, giraffe are subject to a number of ailments including viral diseases (rinderpest, malignant catarrhal fever, cutaneous vital papillomas and lumpy skin disease); bacterial diseases (salmonellosis, anthrax, colibacillosis and tuberculosis); parasitic diseases (of the liver and gastrointestinal tract, for example); and non-infectious pathogens (such as dental and hoof problems, nephritis and atherosclerosis) (Bush, 2003). (The Appendix gives references for many parasites and pathogens that affect giraffe.) Here we will discuss mainly ailments related to the food given to captive giraffe; this is an ongoing concern because in time a poor diet can cause sickness or even death.

To assess the status of nutrition-related health of giraffe at 14 European Zoos, Marcus Clauss, Paul Rose, Jurgen Hummel and Jean-Michel Hatt (2006) analysed 83 of their necropsy reports for animals aged 6 months to a female over 30 years old. Comparing the data in these reports was difficult, because they varied from a short comment on a major finding to detailed reports covering every organ system; nor were the ages of the giraffe always given. (Obviously a complete necropsy report should always be written up upon the death of any giraffe, although some data may be lacking, such as the age of wild-caught animals.)

The reports noted the number of giraffe with various food-related ailments:

- 28 with either no fat, or serous, gelatinous or atrophied fat tissue;
- 8 with excessive tooth wear or dental disease;
- 4 with phytobezoars (indigestible plant material such as cellulose which can obstruct digestion) in the rumen or omasum linked to irregular tooth wear and the eating of grass hay;
- 2 with rumen acidosis (pH of 5.5) of four cases where pH of rumen contents was measured (in 79 cases there was no mention of this parameter); and
- 2 with foreign body ingestion.

Other ailments possibly connected to captivity (because of poor diet, restricted movement etc.) were:

- 15 with arthrosis/arthritis;
- 7 with hoof problems;
- 6 with traumatic fractures of the skull or neck;
- 4 with massive helminth parasite burdens;
- 3 with tuberculosis;
- 3 with pneumonia;
- 2 with listeriosis (see Cranfield *et al.*, 1985 regarding this disease).

Other health issues suspected to be caused by a poor diet include chronic wasting, cold stress, pica, energy malnutrition, inverse serum calcium and phosphorus levels, pancreatic disease, rickets (cured with Ca/P and vitamin dietary supplements), neonatal health issues and urolithiasis ('stones' in the urinary system) (Schmidt *et al.*, 2007, 2009; Ball, 2010; Valdes and Schlegel, 2012).

Urolithiasis

Urolithiasis can be fatal (Wolfe *et al.*, 2000), so it is worrisome that Barbara Wolfe felt its incidence seemed to be greater in the past 10 years than in the 10 years before that. She wondered what changes in feeding regimes might have been responsible for this (deMaar, 2009). To learn more about the problem and its possible relationship with the diet of giraffe living in North American zoos, Kathleen Sullivan and colleagues (2010) sent a questionnaire to 98 facilities to which 41 responded. These zoos, which housed 218 giraffe of various races, were asked to submit samples of feeds, serum and, if possible, urine.

For all the zoos, concentrate and alfalfa hay comprised the primary foods, supplied in varying ratios, with 65% also feeding on browse of 43 different plant species. Only 6 zoos reported a history of urolithiasis in their giraffe (with 7 noting wasting syndrome and 10 sudden deaths). Concentrations of minerals in the diets seemed to be high in general in these affected giraffe, especially phosphorus, and these plus a high urine pH could be contributing factors in urolith development.

Serous Fat Atrophy

As the above necropsy list indicates, health problems related to fat tissue were the most common. The generic name for this problem, Serous Fat Atrophy (SFA), reflects the inadequate diet of many zoo giraffe. SFA is correlated to abnormal fat storage, dental attrition and the feeding of grass hays (leading to rumen content stratification) (EAZA, 2006). To document this, samples of forage and concentrate rations given to giraffe can be analysed in a laboratory to determine amounts of dry matter (DM), crude protein (CP), acid detergent fibre (ADF) and metabolizable energy (ME) (Rose, 2005). Actual foods can be measured on site, including forage, concentrate, fruit and vegetables. The basal metabolic rate (BMR) and metabolic body weight (MBW) can then be calculated for each giraffe from these measurements, taking into account its weight. CP, ADF and ME values vary at different zoos. So does browse intake, which should be essential.

Marcus Clauss and his colleagues (2001) in their studies at the Whipsnade Wild Animal Park of food digestibility worried that giraffe may suffer from SFA because they eat too little. They may not have enough time to eat the food they need, as browsing ruminants they may have trouble handling cellulose-rich fibre structures in their rumens or maybe the food doesn't taste very good – the authors recommend linseed extraction chips and browse as palatable diet supplements. The four Whipsnade giraffe had slightly more protein in their diet than did wild giraffe, and slightly less fat, but the main difference was in the fibre fraction. Zoo giraffe consume much more cellulose (as is typical for grazers) while wild giraffe have a higher intake of hemicellulose and indigestible lignin.

All too often in the past, giraffe have collapsed and died in zoos without seeming to have been sick. A substandard diet was thought to cause pathological changes that led to this 'peracute mortality syndrome', especially where there was added stress such as cold weather. However, the 'acute' is a misnomer as it indicates the death is accidental, when rather a giraffe's faltering condition has been present for weeks or months. The syndrome is characterized by sudden death or by a short illness followed by rapid deterioration and demise. In the above necropsy survey (Clauss et al., 2006), 77% of the 56 giraffe whose age was known had died before they were 15 years old; 48% were in poor or emaciated condition; and many had shown an absence or reduction of fat, or fat in a gelatinous state, or SFA in the coronary groove (on the outer surface of the heart marking the division between the atria and the ventricles), mesentery or in generalized areas of the body (Clauss *et al.*, 2006). SFA is dangerous and widespread in temperate climates.

The role of weather in this discussion came to the forefront in New Zealand in 2005, when John Potter and Marcus Clauss (2005) reported that over an 8-year period, five giraffe had died peracutely, all during the winter. The animals had been maintained in an outside enclosure, moving under shelter (with little heating) only at night. After their deaths, each was found to have had SFA although they had been fed, it was thought, good quantities of browse, alfalfa hay and commercial supplements. However, during the winters the average temperatures at the time of the deaths were 6°C overnight and 15°C in the daytime.

To prevent further giraffe from dying after the fifth death, the zoo rethought its diet and energy intake regime. Because of this, the weights of the two remaining females increased by about 100 kg each during the next 9 months, to 704 and 858 kg; when their calves were born the following November, they were both healthier than the females' previous infants had been. Giraffe must have high-energy sources of food, but they must provide useable energy. A diet rich in starch is a high-energy source, but it causes acidosis which changes the gut bacteria so that giraffe cannot digest the food or fulfil their energy demands in cold weather. Giraffe should be fed food with available energy (so microbes in their gut work efficiently) and lots of structural fibre (ADF/NDF) to provide a good substrate for microbial activity which maintains a healthy rumen (Rose and Clauss, 2006).

Three giraffe died at the Vancouver Zoo in late autumn of 2011 and 2012 when temperatures were quite cold (Crawford, 2012), and one during a recent winter at the Belgrade Zoo, although he had had no symptoms of illness in the previous days. However, after an autopsy was performed, he was thought to have died rather as the result of septicaemia and endotoxaemia caused by clostridial infection (Medi, 2012).

Intestinal parasites from grazing

The Lion Country Safari (LCS) at Loxahatchee, Florida, had an unusual problem: their giraffe had become so relaxed in the warm, humid climate that they liked to lie down for much of the day, often nibbling on the grass around them as they did so. They had become infected with *Haemonchus contortus* larvae (L-3 stage stomach nematodes) that had crawled up blades of grass and been eaten. Once in a giraffe's abomasum or fourth stomach these larvae mature through the L-4 to the L-5 or adult stage. The females feed on blood in their new home, breed, and each releases every day about 10,000 eggs which are excreted in their host's faeces. These eggs then develop to the L-3 stage where the larvae crawl up blades of grass, waiting to be ingested. Two giraffe had died in part because of these worms, while another was successfully treated (Garretson *et al.*, 2009). The use of copper oxide wire particles for intestinal parasites such as *Haemonchus* sp. has also proved effective (Kinney *et al.*, 2008).

The new habit of giraffe grazing and therefore ingesting these parasites was a problem that LCS had to address. Ashleigh Kandrac (2011) describes what is being done to curtail grazing:

- puzzle feeders and more browse are hung high to attract the giraffe;
- when they become bored with the crackers fed to them by visitors, browse is offered instead to lure them back to the feeding platform;
- their enclosure is now enlarged to include a forested area where there are no parasites; and
- keepers increase training sessions to keep the giraffe occupied.

Efforts are also made to curb the parasites:

- donkeys and zebras are introduced to the giraffe pastures each night. These eat the grass and the parasites, but the larvae cannot develop or breed in equid stomachs so they soon die. These grazers also help to keep the grass short which exposes the parasites to direct sunlight, to their detriment;
- a pasture vacuum pulled by a tractor removes manure from areas that the giraffe frequent;
- faeces are collected and analysed periodically to check for this parasite; and
- if an animal is infected, it is removed from the pasture and its pals, and treated with moxidectin to kill the parasites.

Pasture rotation, burns and a winter freeze are known to decontaminate pastures, but these methods are not feasible in a subtropical climate. Zoos in the southern United States find that of all their species of animals, giraffe were among the most affected by gastrointestinal parasites (Kinney-Moscona, 2009). Parasite levels should be routinely monitored (by faecal egg and/or larval counts) and the giraffe treated appropriately.

Fortuitously, foot-and-mouth disease seems not to be a problem: in a few experiments SAT-1 and SAT-2 of the virus causing this disease were injected into giraffe, but there was no dissemination of the disease among the animals (Vosloo *et al.*, 2011).

Locomotory problems

Jurgen Hummel and colleagues (2006) surveyed 74 zoo groups, holding about 350 giraffe, that were involved with the European Endangered Species Programme (EEP). They wanted to determine how many of the zoos had giraffe with leg problems. Most of them (40 of the 74) did: 47% had at least one case of overgrown hooves, 14% had at least one case of laminitis and 35% had at least one case of a joint problem in their giraffe herds. Problems of the locomotory system were similar in males and females; occurred in 63% of the 19 reticulated giraffe and 47% of the 19 Rothschild's giraffe; and 57% of west European giraffe groups were affected, while the rate in eastern zoos was only 35%. Observations included:

- overgrown hooves and laminitis were rare in young giraffe, but more common in those over 8 years of age which carry far more weight;
- these two ailments occurred where indoor space was small (26 m^2 per animal) more often than where indoor space was larger (42 m^2 per animal);
- these two ailments had no correlation with diet; however, grazers in captivity seldom had hoof problems unlike browsers such as moose, okapi and giraffe, indicating that for them there may be a connection between diet and hoof problems (Clauss *et al.*, 2003);
- joint problems were more common in the front legs of giraffe (24 cases), with 5 cases in the hind legs (3 of them caused by inbreeding or trauma) and 5 in both front and back legs; and
- giraffe with laminitis were more likely than others to have more easily digested food such as bread, pure grains, fruits and vegetables in their diet.

The authors recommend that if giraffe have foot problems, a shift from a hard floor such as concrete to a varied one that has some hard areas with abrasiveness and some sections of soft flooring might be a good idea. They also give advice on bedding material, temperature of indoor areas, humidity and nutrition. Anke Egglebusch (2008) agrees that hoof and joint problems depend not on a single factor, but on interactions of floor substrate, weight/size, nutrition, heredity and movement level. Movement level itself is associated with time of exercise, size of enclosure, group composition (sex, age, other species) and individual interactions. Other hoof problems are discussed below.

The very length of their legs and the heavy body they support make giraffe vulnerable to leg problems, as has been found in many zoos. However, there are now various ways to repair the legs to full use again, as the following examples indicate.

- A 4-month-old captive reticulated male hurt his right hind leg when he was frightened, ran into a fence and became lame. Radiographs showed that he had fractured part of the femur. Under anaesthesia bone fragments were removed and the injury repaired so that after 6 months he could walk and run with only minimal lameness (Quesada *et al.*, 2011).
- A 20-month reticulated giraffe at the Münchener Tierpark Hellabrun had bilateral upward fixation of the patella corrected with two surgeries under anaesthesia (Kempter *et al.*, 2009).
- An 8-month-old reticulated giraffe developed extreme lameness in his right front leg, with swelling of the metacarpophalangeal joint, presumably because of a traumatic accident. With the use of an arthroscope to examine the area, synovitis, cartilage damage and an osteochondral fragment were identified and the damage repaired. Arthroscopy in a giraffe had never before been reported (Radcliffe *et al.*, 1999).
- A 10-year-old Masai giraffe developed an acute lameness in his right front leg which made it painful for him to walk. Radiographs revealed a fracture of the medial claw of the distal phalanx penetrating into the distal interphalangeal joint (James *et al.*, 2000). The giraffe was sedated when standing in a chute and a wooden 'hoof' attached to the lateral claw of the same limb. Within 3 days the animal was no longer lame. The wooden block which was made of pine wore down completely after 2 months, before the fracture itself was completely healed.

- A young reticulated male who had swollen hock and fetlock joints was diagnosed with a mycoplasma-associated polyarthritis because a *Mycoplasma* sp. was eventually detected in his synovial fluid (Hammond *et al.*, 2003). For four recurrent bouts of lameness he was treated at first with antimicrobial therapy, but in each case had a relapse after about 2 weeks. He was finally cured with an oral enrofloxacin treatment.
- At Cameron Park Zoo in Waco, Texas, a female called Julie with a right front fetlock defect was able to manoeuvre successfully for most of her life with a prosthetic shoe. Her condition worsened by age 20, when she was offered further extensive support by Rachel Chappell and Krista Seeburger (2012). They won an award in 2012 for their devoted care of her from the International Association of Giraffe Care Professionals.
- At the Oakland Zoo in California in 2001, a young male suffered a trauma-induced septic arthritis of the fetlock joint (Phelps *et al.*, 2012). Treatment included surgery in 2004 for arthrodesis of the fetlock joint. This caused a deformation of the cannon bone above the fused fetlock, making the two forelimbs different lengths. However, with an extensive operant conditioning programme, he continues to thrive.
- A newborn female at the Guadalajara Zoo had flexor tendon laxity in all her legs so that she was unable to stand up or walk; this seemed to be a genetic condition. However, this giraffe was hand-reared and cared for so carefully by the staff that she recovered completely (Cárcoba, 2012).
- A female reticulated giraffe born of a father–daughter cross who displayed hind limb lameness was found later to be suffering from hypothyroidism (Mainka *et al.*, 1989). Other ailments included hoof and skin abnormalities, anaemia, an atypical hair coat and a poor reproductive history.
- Giraffe, like people, may be subject to osteoarthritis. This happened to a 16-year-old female Rothschild's giraffe who limped because of her affected limb joints (Cracknell *et al.*, 2010). She was treated with phenylbutazone for 3 years, but then began losing weight, so was euthanized on welfare grounds. After dying from the anaesthetic, a team of scientists immediately took the opportunity to insert a left flank laparoscope into her abdominal cavity where they were able to look at a few of the organs there despite crowding from the intestines and rumen. They report that laparoscopy could be a useful tool in the future for the treatment of giraffe.

Hooves

Hooves* remain healthy in the wild where giraffe can wander immense distances in search of food or sex. They are challenging in zoos, however, where there is little room to walk about and they may grow abnormally because of diet or substrate. When giraffe are kept on grassy areas rather than more abrasive ones such as concrete, their hooves may grow too long and have to be trimmed. At Longleat, in England, the hooves of one male have been trimmed four times successfully, but one giraffe in poor health died because of the stress of this operation (Cawley, 1975). There is some evidence from the Longleat giraffe that hoof growth might be hereditary because if parents were affected, often their offspring were too; of the 17 giraffe at the reserve, 5 had hoof abnormalities.

Often a giraffe can be trained to enter a chute and stand still while a keeper trims his or her hooves. Sometimes, however, it has to be anaesthetized, for which there are two different protocols: medetomidine/ketamine or xylazine then etorphine (Lécu, 2010). When the giraffe is recumbent, its four legs should be positioned for easy access to its hooves and for later ease in enabling it to stand after the operation. Using radiograph pattern measurements, each hoof can be reshaped using electric grinders and other tools. This should be done quickly but carefully, because variations in hoof angles and lengths will cause changes in ligaments, foot support and gaits which may prove to be painful. Reversal antidotes should be administered mainly through the veins to speed up recovery. Trimming the hooves of a giraffe must be taken seriously as it can be lethal (Cooper, 2010).

Stereotypical behaviours

In zoos, behavioural abnormalities have also been linked with food and feeding problems (Baxter and Plowman, 2001). Because their feeding time each day is much reduced in zoos compared to in the wild where they move constantly to find food, many animals in zoos develop stereotypical behaviours defined as invariant patterns which are regularly repeated yet serve no apparent purpose (Mason, 1991). Captive giraffe are no exception, living as they do in small enclosures with food that is foreign to them. Meredith Bashaw and three colleagues connected with Zoo Atlanta (2001) carried out a survey of such behaviours in 214 giraffe and 29 okapi resident in 49 North American zoos; most of these animals (66%) were females and all but 2% had been born in captivity. By far the two most common stereotypes were the licking of non-food objects such as outdoor fences reported for 72% of the animals, and pacing which occurred in 29%. A few individuals (3%) exhibited stereotypes of self-injury, head tossing and tongue playing.

Stereotypies seem to be related to highly motivated behaviour patterns, one of which is feeding; in zoos this involves new diets, often concentrated foods, new presentations of the food and the short time needed to consume it. In the wild, giraffe spend much of the day, especially in the winter or dry season, browsing. Consuming leaves from acacia thorn trees, one of their favourite foods, involves carefully lifting leaves away from thorns with their tongue. By contrast, food comes too easily in zoos – high-quality food such as alfalfa which requires little time to consume may increase oral stereotyping. Similarly, giraffe in the wild walk for miles each day, so curtailing this activity as occurs in zoos, especially in indoor paddocks, induces stereotypic pacing.

The data analysed from the 49 zoos found that animals were significantly less likely to show stereotypical licking if they:

- belonged to the Masai or reticulated subspecies;
- spent more time housed indoors;
- lacked access at night to members of the same species;
- were fed only by staff;
- received browse; and

- had access to feeders with closed tops which took more time than those with open tops to circumnavigate.

Because three of these factors involve feeding, licking seems to be related to a feeding motivation.

In the analysis of pacing behaviour, the researchers found that giraffe were less likely to do so if they:

- were of an unknown or Masai race;
- lived in a larger indoor enclosure;
- had experienced an environmental change in the past year; and
- were not fed concentrated chow.

Pacing seemed most closely correlated with environmental factors, especially having too little space in which to roam. With reduced opportunity to exercise, giraffe may sleep less and so have increased time available for undesirable behaviours, especially in indoor enclosures where there is limited stimulation and possible overcrowding.

Judging from the results of the survey, if one were planning ideal housing for captive giraffe it would be best to have as large an enclosure as possible, a mineral lick and much browse with fibre available, use of closed-top feeders, one staff member dedicated to supplying food and some variation in where food, toys and such are located. The idea is to increase the time giraffe spend feeding by increasing quantity, processing-time and distribution range of the food.

In an effort by Zoo Atlanta to reduce stereotypical licking, fence areas that attracted licking from the giraffe, Betunia, were sprayed with Grannick's Bitter Apple, a harmless but bitter chemical (Tarou *et al.*, 2003). Betunia was aware of this substance because she reduced her licking where it was sprayed, but she continued licking at unsprayed areas of fence, so the experiment was considered a flop.

Betunia became involved in new experiments, however, to determine if licking behaviour could be curtailed or stopped by giving her and the other Zoo Atlanta giraffe, Kamili and Aaron, fodder in containers that made it more difficult for them to obtain the food (Fernandez *et al.*, 2008). The aim was to overcome 'the inability of a highly motivated feeding behaviour pattern, tongue manipulation, to be successfully completed' (p. 200). Ingeniously, Atlanta personnel devised nine types of feeders, all of which forced the animals to some extent to slow down their intake of grain or alfalfa; holes, if present, were only big enough to get their tongue inside but not their muzzle. Each type of feeder was either stationary or allowed to swing free high in the air.

She had been the most obsessive licker, but Betunia reduced this behaviour significantly as she fed from the increasingly complex feeders, although this successful progress was less clear-cut for Kamili and Aaron. For all three giraffe, mesh feeders in which large metal 'jacks' mixed with the food, emulating thorns on acacias around which the giraffe tongues had to navigate, were effective in increasing the length of rumination, if not in reduced licking. The experimenters conclude that licking of non-food items can be reduced by having giraffe forage in more naturalistic ways, but they were unable to eliminate it completely.

Figure 10.1   Puzzle bucket used at the Brookfield Zoo so that hungry giraffe have increased tongue exercise to try to prevent stereotypies. The 'jack' is put with food into a hanging 5-gallon plastic bucket which has holes cut into it that will accommodate a giraffe's tongue but not its mouth. Photo from Chicago Zoological Society/Schulz/Amy Roberts.

Maybe stereotypies are more correlated with rumination than with feeding? To test this, E. Baxter and A. B. Plowman (2001) experimented with two giraffe at Paignton Zoo Environmental Park in Devon, an 8-year-old male and a 22-year-old female. For 4 h each day, for 12 days, the researchers recorded the amount of time these giraffe fed, ruminated and exhibited oral stereotypies which included licking walls, bars, trees etc., and tongue playing by thrashing their tongue from side to side in the air. Then, for 6 more days, they added coarse-cut meadow hay to their diet while continuing their 4-h daily observations.

The results were significant. Both animals increased their rumination time when they ate the coarse hay each day, by 9% for the male and 12% for the female. Its addition did not have a significant effect on the time spent feeding, but did significantly reduce the time spent performing oral stereotypies which were far more frequent in the female (who had very worn teeth) than in the male. Often she ruminated and licked at the same time. Having meadow hay in their diet also regularized the timing of their behaviour; they now had distinct dawn and dusk feeding peaks and longer, more regular bouts of chewing their cud during the middle of the day and overnight. The authors concluded that despite the small sample size, a decrease in oral stereotypies seems to be correlated through the effect of diet on rumination rather than on feeding itself. This indicates that providing coarse hay to the diet of giraffe is an inexpensive way to increase rumination and reduce oral stereotypies, thus making these behaviours more like those of wild giraffe, the gold standard.

At the Twycross Zoo in England, efforts to prevent pacing and stereotypical behaviour involved feeding branches of 16 species of local trees to the three giraffe, about 15 bundles at intervals during the day and in the night enclosure. Bark and leaves from willow and cherry branches were quickly eaten, while bundles of mature hawthorn branches provided thicker bark and largish thorns which kept them occupied longer. Stinging nettles (*Urtica dioica*), garlic and onions mixed with lucerne stuffed into a metal mesh ball provided giraffe with a nutritious treat as well as sensory stimuli as they wrestled with their tongue to free the food. High-fibre, low sugar 'chaff' was also added to meal rations to bulk them up (Rose, pers. comm., 2012).

As usual, experiments often answer some questions but raise others (Rose *et al.*, 2008; Rose and Roffe, 2010).

- In some ungulates, high-caloric diets of concentrated foods produce excess acid in the gut. This has been linked to oral stereotypy in other ungulates. Maybe licking persists because it increases the production of saliva which may buffer gut acidity?
- Stereotypies have been correlated with stress, so perhaps a reduction of stress can reduce these aberrant behaviours?

Sonja Schmucker, Lydia Kolter and Gunther Nogge (2010) have studied tongue movements using video recordings during browsing, licking and tongue playing (the last two referred to as 'oral disturbances') by giraffe in the Cologne Zoo in Germany. The four adult females had oral disturbance frequencies varying from 4.6% to 30.7% per winter day. Such activities decreased significantly in summer when the animals were fed on browse. Browse or hay held in closed narrow-mashed baskets increased tongue movements significantly in any season. In general, different foods and their varied presentations provided varying amounts of oropharyngeal stimuli. The average number of total tongue movements per day per giraffe was the same in different seasons, but normal feeding contributed significantly less in winter when oral disturbances were greatest than in summer.

Enrichment programmes

The aim of progressive zoos is to make conditions for their animals as much like those in their wild habitats as possible, and to encourage natural foraging and the social behaviours that occupy their days in the wild. Enrichment methods to help simulate wild conditions are constantly being updated and invented to engage different sets of behaviours and arouse different senses in each species. In this way the animals are less likely to become bored and to fall into negative stereotypical behaviours. Stimulation can be addressed by nutritional, sensory, physical, occupation and social stimulatory tools (Rose and Roffe, 2007).

The Oakland Zoo takes great care when planning enrichment programmes for the giraffe (Porterfield, 2012), considering all possible dangers that might lead to injuries such as strangulation, head hooking, ingestion, choking, tripping, entrapment, facial scraping, cutting or slipping. Giraffe are liable to panic, so each new innovation or device is introduced slowly. Successful innovations (described in detail by Jessica Porterfield) can be fashioned by hand.

(a) Fasten browse bunches together in a holder on a fence so that the branches stand up in a natural style rather than down. Giraffe enjoy not only the browse, but putting their heads right among the leaves (Figure 10.2).

(b) Put treats suspended in a paper bag over the feeder so that a giraffe can tear open the bag to get at the food.

(c) Make chimes with lengths of dry bamboo hanging down in a bunch so they touch each other in the wind, making a sound. Some giraffe were afraid of this at first, but then they enjoyed pushing it with their heads and smelling it. Food bits or scent can be placed near the chimes (Figure 10.3).

(d) Hang up high a length of wide pipe with small holes in it. Concentrated scents can be put in this. Use a variety of smells to see which are enjoyed by which individual.

(e) Hang up coarse mats on the fence at different heights so giraffe can rub against them to scratch themselves (Figure 10.4).

(f) Introduce a new surprise into the giraffe paddock every few days to attract their attention – a figurine, a large ball or an article of clothing.

At the Dallas Zoo, zoo keepers addressed the enrichment needs of their large herd of giraffe ranging in age from 1 to 23 years (Six and Streater-Nunn, 2012). They installed a raised 'arm' from which to hang enrichment items at different heights, focussing on objects that developed play and tongue work and that 'invited extended periods of interaction and allowed for usage by multiple individuals'.

At Twycross Zoo in England, Paul Rose and Sarah Roffe (2007) tested lavender oil (*Lavandula angustifolia*) to see if its smell interested the giraffe as it does some other species. They hung the oil mixed with barley straw in the giraffe house where it could be smelled but not eaten. The giraffe continued to chew their cud in its presence, which was a good thing, but fed and moved about less than before, and, worryingly, it increased stereotypical behaviour such as pacing.

Training programmes

Giraffe are big and strong, but also highly nervous, timorous animals who must be treated sensitively in zoos to enable them to overcome their fears. Recently in the Bronx Zoo a zebra kicked a blue ball over the fence into the giraffe enclosure; appalled, the giraffe refused to come out of the Giraffe House for the rest of the week (Samuels, 2012). Helping them cope with zoo life can be done with careful training which can prevent a giraffe from panicking and running amok, hurting itself or others.

Why else should one train giraffe? There are a number of reasons (Phelps and Clifton-Bumpass, 2011; Bru, 2012):

● most importantly, too often giraffe die when anaesthetized, so if this procedure can be avoided it will save lives;

● training creates a bond of trust between giraffe and staff, making both their lives easier – a giraffe will readily follow a trainer staff member into the barn, for example;

Figure 10.2   Browse holders are favoured by giraffe at the Oakland Zoo, all of whom are reticulated.
Photo by Jessica Porterfield.

Figure 10.3   Mabusu at the Oakland Zoo listens to the bamboo wind chimes. Photo by Jason Loy.

Figure 10.4   Benghazi rubs his nose on a scratching mat at the Oakland Zoo. Photo by Jason Loy.

Figure 10.5   Twiga at the Oakland Zoo, about 9 months pregnant, communes with a *papier mâché* model. Photo by Erik Beckman.

- individuals learn to respond to their names and become known individually;
- giraffe can learn how to paint, to produce saleable drawings (I have mine over my desk) (Figure 10.6);
- the taking of blood and small medical treatments are far less stressful for a trained giraffe (Martinez, 2010); and
- trained giraffe are great ambassadors for the species; they interact well with the public and therefore star in conservation programmes.

Training such as that proposed by Felinova Animal Behaviour Consulting, and described in some detail by Anneleen Bru (2012), involves a clicker and positive reinforcement by way of treats. Whenever a staff member uses his or her clicker in response to a positive giraffe behaviour, the animal undergoing training earns a treat, such as a piece of banana. With a training session of 15–20 min given 3–5 times a week, a giraffe can be trained in as little as several weeks to do what is wanted in return for the treat. Staff members kept notes on each training session of each animal including in their records the object of the exercise, location, presence of people, treat given, success of the session and suggestions for future trainings.

An example of the usefulness of positive reinforcement training occurred when the end of a female giraffe's tail somehow became degloved, exposing six inches of tail vertebrae. Because she had been trained, keepers could keep her standing calmly inside a restraining device where she could be injected, touched and have her tail docked (Peterson, 2012).

Figure 10.6   At the Oakland Zoo, Benghazi paints a picture. Photo by Amy Phelps.

Figure 10.7   'Look at the board please' – training Khalid to do as asked at the Oakland Zoo. Photo by Amy Phelps.

Amy Phelps and Lisa Clifton-Bumpass (2009) have developed a sophisticated training programme for giraffe at the Oakland Zoo in California to address their general nervousness and fright/flight response to all new stimuli. This involved the following, for example.

- Conditioning individuals to a variety of new objects to sniff or touch or somehow interact with in a game called '101 things to do with a box'. The giraffe is presented at each training session with such things as stuffed toys, traffic cones, a ball, a piece of clothing or a hula hoop on the ground to put a foot into. Eventually, if an animal's hooves have to be trimmed, this can be safely done with the giraffe's cooperation learned from the hula hoop.
- Presenting the giraffe with the 'Stranger is not a Danger' game in which it is introduced at each session to an unfamiliar person rather than an object. The person keeps his or her distance until the giraffe relaxes, then moves slowly nearer a few small steps at a time. People sometimes dress as veterinarians to familiarize the giraffe with this clothing.
- Gradually, a giraffe also learns to respond to his or her name,[2] and to 'move up' or 'step back' or turn this way or that if a medical procedure is necessary. A well-trained giraffe

---

[2] The giraffe in the Bronx Zoo have all been named either Margaret or James, depending on their sex, based on the terms of the bequest by the benefactors, James and Margaret Carter, who support the animals. Fortunately, they were each also given a second name to which they respond.

Figure 10.8   Team training to teach T'Keyah to lift her leg involves a feeder giving treats, a trainer moving the leg and a coach to coordinate their activities. Photo by Lisa Clifton-Bumpass.

will open its mouth so its teeth can be carefully examined. Even acupuncture for pain has been applied to a giraffe's skin without it being unduly upset.

The original training began in 2001 when Phelps and her colleagues were caring for a reticulated male with a traumatic septic arthritis of the fetlock joint. Since then, they have greatly expanded the training programme which now includes techniques to deliver intensive hoof care, chiropractic care, massage therapy, IV antibiotic therapy, drawing of blood, injecting drugs and applying transdermal fentanyl patches for pain management. Phelps and Clifton-Bumpass have delivered a training programme at other facilities based on the operant conditioning techniques noted above that emphasize trainers working together as a team and giraffe competently cared for and gaining confidence in their zoos (Cotter *et al.*, 2012). Training is now being implemented as far away as the Uganda Wildlife Education Centre (Opio, 2012).

Extensive training has also been carried out successfully at Fort Wayne Children's Zoo, instigated initially by the need to shift a herd of seven giraffe from an old barn to a new facility at the opposite end of the giraffe exhibition space. Aimee Nelson (2012) describes in detail the infinite patience necessary to deal with nervous individuals. For example, Kali, who arrived at the zoo at age 15, was so cautious that she had refused for 5 years to follow the other giraffe each day to the exhibition area. Giraffe are always fearful when faced with new challenges, but because of the training and extensive planning the move was completely successful. In the new quarters there is room for further training by positive reinforcement; already staff can do voluntary blood draws, hoof care, radiograph training, injection training and eye/ear examinations.

At Fresno's Chaffee Zoo, Kit Perry (2012) reports that training has gone further, in that giraffe at this facility are taught to respond to what is a 'visual name'. Vision is the giraffe's primary sense, so each animal is given a unique colour and shape target. Then he or she is trained to come to that target for a treat, but to ignore targets of other individuals for which they receive no treat. For example, Uzuri would ignore her name as usual when someone called her by that name, but moved quickly to her visual name, a yellow diamond, when a keeper held up that symbol. Perry reports that this training to a target works well when:

- a keeper wants to examine a giraffe that may have hurt itself; he or she can call the animal over by holding up the target without disturbing the other giraffe;
- if giraffe need to be separated and/or grouped for exhibition, each animal can be summoned to a particular area; and
- a mother and calf can be drawn into the shade on a hot afternoon without other animals being involved.

These animals have also undergone traditional individual training during which they have learned to follow a keeper, come here, back up and bend down. The last command which has a giraffe lean down to touch their target on the ground is used to show the public how a giraffe drinks water in the wild. At the Twycross Zoo in England, target training sessions are affected by prevailing weather conditions, but apparently not by time of day or by crowd size (Rose and Roffe, 2010).

Figure 10.9 Amy Phelps exercises the leg of old Tiki as part of her physical therapy programme. Photo by Anne Dagg.

Figure 10.10 The author bonds with the reticulated giraffe Zuri at the visitor feeding platform of the Phoenix Zoo. Photo by Anne Dagg.

Visitor feeding programmes

Once giraffe have been trained by their keepers to do a variety of things, they can be taught to accept food from strangers. Feeding giraffe is a welcome new activity. It is unknown in the wild, of course, but in zoos it is becoming increasingly popular for a number of reasons:

- it offers diversion and stimulation for the giraffe after they are trained to accept the food offered;
- it is a thrill and a learning experience for the visitors who offer them food – they can examine a giraffe's head close up and notice how some individuals are brave in coming forward while others are less so (Porterfield, 2012);
- visitors pay money for this privilege to the zoos, which always need this;
- it may spark interest in conservation issues because the giraffe is a symbol of species that could actually become extinct in the wild;
- volunteers themselves can be trained to oversee the feeding as at the Phoenix Zoo (Schultz *et al.*, 2010), thus incurring minimal expense for the zoo.

A possible downside, however, found by David Orban and his team (2012) at various American zoos, is that giraffe who spend time at the feeding stations are also more idle than other giraffe. This study is ongoing.

When a visitor feeding programme was first instituted at the Fort Wayne Children's Zoo, the two key issues were that the feeding platform be safe for both giraffe and visitors, and that the food be nutritionally appropriate, palatable to the giraffe and ingestible in large quantities (Eagleson, 2010). These matters were satisfactorily arranged, and have resulted in a thrill for over 612,000 visitors, some very well-fed giraffe and a less cash-strapped zoo.

Eric Dunning (2010) of the Cincinnati Zoo describes the training of their Masai giraffe for its new feeding programme and the selection by trial and error of their favourite foods. All five giraffe seem now to be happily involved in the visitor sessions.

Rarely, a person's health may be affected by a giraffe. A young woman who voluntarily helped look after one at a zoo developed hives on her forearms where they had contact with the animal's fur (Herzinger *et al.*, 2005). Skin prick tests found that she suffered from 'an "exotic" immediate-type contact allergy to giraffe hair'.

Social attachments

One important feature in the life of captive giraffe is that their feelings for each other cannot be manipulated by the zoo. They have to eat what they are given, live in small quarters, be subject in most things to their keepers, but they can make their own friends.

What do animals think about each other? How and why do they interact with the individuals among whom they live? People who study animal behaviour want to know such things. I certainly wanted to learn about giraffe activities which might tell me how they understand their world. I was thrilled when, on my husband's sabbatical in Sydney, Australia, I was able to spend each day at the Taronga Zoo with my toddler daughter.

Over a 5-month period I sat outside the giraffe paddock noting down every time one giraffe touched another (Dagg, 1970). This seemed like a clear-cut way to help clarify their feelings toward each other.

The herd of 18 giraffe comprised three adult males – the 24-year-old male Jan and two males in their prime: Oygle (7.5 years) and Lumpy (4.5 years); 11 females including one, Cheeky (11 years), who had never been pregnant; and 4 calves all under 6 months of age. My notes, accumulated over 57 h, yielded information on 1447 encounters. After analysing these data I came up with a list of statistically significant behaviours. In order of frequency, results showed:

- old male Jan no longer interacted much with the other giraffe, although he had earlier fathered a number of calves;
- the four calves were interested in each other, but not much in the adults;
- males often sparred with and hit each other, but they hit Cheeky much harder than their male pals. Their blows at Cheeky were often violent; once she ducked as Oygle swung at her so that he hit the fence instead with his head, making it bleed profusely; and
- nosing, licking, rubbing against, sniffing genitalia and flehmening were mostly done by the two prime males. The mothers nosed and licked their own calves in what seemed to be bonding behaviours.

These giraffe also participated in various activities rarely or never seen in the wild (underlining the importance of always keeping data collected from wild and captive animals separate):

- males often nosed, licked and rubbed against each other; such attentions were astonishing, given that breeding adult males do not spend time together in the wild;
- mothers suckled their calves about every 5 h, but when a calf went to its mother to suck, often other giraffe rushed over to suck too. Once four giraffe, some adult, sucked at the same time. Sucking sessions never lasted longer than a few minutes;
- females sometimes flehmened or licked male urine;
- males hit a female (Cheeky) who hit back, but not nearly as often or as powerfully;
- females never mounted any giraffe in the wild, but a female calf mounted other calves four times. The males did not mount each other, although such homosexual behaviour is common in the wild where there are many subadult males in an area (Innis, 1958);
- the male giraffe were never truly aggressive toward each other (as they were against Cheeky);
- giraffe shared 191 times with another the leaves on a branch it was holding in its mouth. By contrast, for eight other species the males had to be separated from each other during the mating season to prevent combat or death. Usually there is no need to castrate a giraffe in a zoo because of combative behaviour, although this can be done (Borkowski *et al.*, 2009).

It is not surprising that strange things happen in zoos that do not happen in the wild, given the unnatural conditions of captivity. In the wild two male adults would never be together for any length of time; they would be mostly alone, wandering about to find females in oestrus with whom to mate. Females in the wild are often in small groups, but the composition of these groups changes frequently, unlike in a zoo. The calves would be used to being together in both venues, but they would be less likely to have regular chances to nurse in the wild. And, because the adults did not have to spend many hours each day hunting for browse and eating leaves, they had, figuratively, time on their hands to do other non-productive things as discussed earlier.

Loraine Tarou, Meredith Bashaw and Terry Maple (2000) at Zoo Atlanta decided to find out more about giraffe behaviour by seeing how two females in a zoo would react when their third cage-mate, a male, was taken away. The two females were of mixed race, Betunia aged 16 years and Kamili aged 11 years. The three had been together for 9 years.

Stimulation for all animals is important in zoos, but too much may cause stress. The two females were upset when their male friend disappeared. Their level of activity increased. Betunia spent more time than before licking the exhibit fence while Kamili began licking objects she had usually ignored, and pacing back and forth in the afternoons. They hung out in a smaller area of their paddock than previously, and spent more time than before standing in contact with each other and rubbing their necks together. This behaviour in the wild is common for subadult males, but not at all for females.

There seemed no doubt that Betunia and Kamili were stressed by the removal of the male they had known for many years, just as species in the wild are upset, especially higher primates, when animals are killed or leave their community. The authors conclude that care must be taken by zoos when they remove animals from their groups because this can have a negative psychological effect on all the individuals (Rose, 2012). Indeed, Paul Rose (2011) suggests that there may be a direct correlation between the number of times a female is moved and the number of youngsters she can produce.

In zoos where giraffe cannot wander far, it is possible for keepers to document special friendships, defined in part as those pairs who spend more time close together than do other pairs. Meredith Bashaw continued her interest in giraffe psychology and with three colleagues (2007) carried out such a study at the San Diego Zoo's Wild Animal Park. Over a 2-year period, she drove early each week-day morning through this area in a vehicle to which the six adult females and other giraffe were habituated, noting down at 1-min intervals which two animals were within two neck lengths of each other, and what they were doing. Often they were feeding at this time at one of 11 feeders; these were spread out so there was no reason why two giraffe were close together if they did not want to be. As well as being near each other, social interactions (in which A and B interact interchangeably) included:

- approach – A directly approaches B;
- necking – A rubs or entwines her neck with B's neck;
- head rub – A rubs her head on B's body (but not on her neck or head);
- bumping – A pushes against B, usually with her chest;
- social exam – A sniffs or licks part of B's body (but not her muzzle or anogenital area);
- muzzle/muzzle – A and B touch or sniff each other's muzzles;
- co-feed – A and B eat at the same plant or branch and so are close together; and
- sentinel – A approaches and stands by B who is lying down.

Figure 10.11   Reticulated male Kayode and an eland get along well in the African Veldt, Oakland Zoo. Photo by Anne Dagg.

The most common affiliative interactions for two animals were co-feeding (50% of 2748 friendly interactions) followed by approach (26%), social examination (11%), necking (6%) and bumping (5%). These activities did not happen randomly between any two pairs, however. Although there were 12 giraffe in all, the notable social relationships involved six *G. c. rothschildi* females. Most interactions occurred between mother–daughter pairs and pairs with large age differences between members. This is reminiscent of the situation in the wild where mothers and their young remain together for a year or more, and where one female may stay with a group of youngsters during the day while their mothers go off to feed elsewhere. This same pattern of a mother and her young remaining close for years is absent in wild herbivores such as caribou, bighorn sheep and mountain goats (Dagg, 2011).

Bashaw (2011) carried out similar research about friendships among giraffe in another facility, Maryland Zoo in Baltimore, to see if they behaved in more or less the same way. They did. The four adult females displayed special friendships among themselves, with in one case a mother closer to one daughter one year, and to another the next. Three sets of female giraffe in zoos in San Diego and Baltimore also had similar activity patterns during the daytime, in order of grouped frequency: feeding, inactivity, ruminating, stereotyping, locomotion, vigilance, social and 'other'. The category 'social' included sparring and mounting, rare activities observed more often among the Baltimore females.

Figure 10.12 Dagmar lies down near the giraffe house at Chester Zoo. Photo by Paul Rose.

Perhaps this was because, having less physical space in which to manouevre, they were always in closer contact with each other?

Paul Rose (2010) carried on research related to that at Zoo Atlanta with further work at Marwell Wildlife in Hampshire, UK. This time friendship itself was the focus among captive giraffe, particularly the stability of special bonds between individuals during three successive years, within populations of 6, 10 and 7 giraffe, respectively. Three undergraduate students collected data (60 h per year) for each animal on her nearest neighbour (within one neck length) over time and the time they spent interacting versus time spent apart.

Although the composition of giraffe in the caged population changed each year, strong relationships remained among the older animals in the group, namely Grace,

Figure 10.13   Paul Rose with Yoda, a Rothschild's giraffe at Paignton Zoo Environmental Park. Photo by Paul Rose.

Matilda, Isabella and Mary, bonds that were stronger here than those between mothers and their calves. Ties between Matilda and Isabella were particularly intense, lasting during the three years. In the second year, however, Mary tended to keep more to herself. These results were different than those found at Zoo Atlanta where the strongest ties were between females of different ages, such as mothers and daughters. Clearly for individual animals, even in changing groups, two may retain a strong preference for each other.

Increasingly, many zoos have large African sections that allow giraffe and other non-predatory species to mingle freely, as they could in their natural habitat. As long as species and individuals are carefully chosen for compatibility, there should be little or no dissension among the animals – indeed, giraffe are far more peaceable than other ungulates during their mating seasons (Dagg, 1970b). In the Oakland Zoo a young adult male giraffe, Mabusu, is especially friendly with a 2-year-old eland, Kairu (Phelps and Mellard, 2012). Rather than be with their own kind, they spend 85% of their time together, sparring gently with their horns, feeding side by side, following each other about, licking and smelling each other's faces and necks, and resting together on the same straw bed. This friendship is a positive and enriching experience for both.

Housing

Giraffe need as much space in zoos as possible, both indoors and out. The outside yard should have some shade and a dense substrate that drains well, but not one made of small crushed stones that might work through the bottom of a hoof (Bush, 2003). Giraffe must be kept indoors (with heating) when the temperature falls below 50°F (10°C) where there should be separate stalls for males and mothers with young, and a restraining area with a chute or gates for minor medical treatments. The flooring should be well drained, with roughened concrete for good footing and proper hoof wear.

Houses for giraffe in zoos must be high, given the unique shape and size of their inhabitants. When Nashville Zoo's barn opened in 2006, modern ideas for the Masai giraffe focussed on the hay room, sand yard, mechanical room, mezzanine loading dock, giraffe loading dock, individual stalls and giraffe restraint device (Meinhardt and Teravskis, 2010). This state-of-the-art facility allows for advanced husbandry practices, varied training opportunities and medical procedures.

The Sacramento Zoo has also recently renovated the facility for its reticulated giraffe, this time building the new facility over the old one which presented unique problems for the giraffe in residence (McCartney, 2010).

11 Status and conservation of giraffe races

Why races?

Ancestors of the giraffe came into Africa from Asia through what is now Ethiopia, then spread out over the continent. When members of *Giraffa camelopardalis* evolved, probably in the Pliocene (Churcher, 1978), they spread throughout Africa wherever there were savannahs and open woodlands; their range was restricted only by a lack of trees and bushes from which to forage (although they did not inhabit dense forests) and by too cold weather in the south. Populations of giraffe became isolated in different areas, and over as much as a million or more years developed along different genetic paths leading to distinct subspecies/races. Nine of these have been commonly accepted for giraffe (Lydekker, 1904; Dagg, 1962a, c; Brown *et al.*, 2007), but as we shall see, there are still grey areas.

As men with guns advanced into the interior of Africa in the 1800s, they shot many of these giraffe and gradually took over the best land to farm and build towns and cities which precluded the presence of large wildlife. The races were all undermined to some extent, and some are now on the verge of extinction. Today, the distribution of giraffe is patchy and discontinuous. Some giraffe have been reintroduced to new areas (notably *G. c. rothschildi*) where it is hoped they will thrive, although occasionally these extralimital areas had never before been inhabited by giraffe, as at the southern edge of the continent.

To preserve all species of wildlife, governments, individuals and non-governmental organizations (NGOs) of countries in Africa have set up protected areas dedicated to native animals and plants where people cannot live, hunt or farm. In Kenya, where tourism is a major industry and home to three races of giraffe and large numbers of other species, David Western, Samantha Russell and Innes Cuthill (2009) carried out a survey comparing data on the status of wildlife living in protected and in non-protected areas. The results were disappointing. Far from being a refuge for animals, the protected areas had been as vulnerable as the non-protected ones; over the past 30 years animal populations had declined sharply, at a similar rate, in both types of regions. The largest parks, such as Tsavo East, Tsavo West and Meru, which might have been thought the safest, were actually the worst off. Their high losses were related to habitat change and the difficulty of protecting animals in large remote areas. Migratory animals were particularly hard to protect, despite Kenya's effective security services – since 1989, poaching has been largely contained, reflecting the growing number of rhinos and elephants, the two most vulnerable species.

A number of recommendations arise from their report:

- ecological factors (such as drought) and human-induced problems should be separated, so the latter can be addressed even if the former cannot;
- wildlife areas should be surveyed often using set parameters in order to detect problems as soon as they arise;
- pastoral and agricultural expansion into some parks (as Tsavo) should be curtailed; and
- each park and reserve should have a habitat management plan which prohibits rangeland burning among other things.

The researchers note that in contrast to nature reserves and national parks, some private and community sanctuaries have stable or increasing wildlife populations because of joint-venture partnerships with private tourism stakeholders. The engagement of communities around Amboseli National Park has fostered tourism and an increase in wildlife, for example (although lions and elephants have been killed) – so such facilities along with their outreach programmes should be encouraged. New policies should be established in Kenya and other countries that combine national, community and private initiatives in order to sustain large free-ranging populations of giraffe and other wildlife. Such ventures have been a huge success in Namibia in increasing wildlife numbers and translocating giraffe into communal areas (Fennessy, pers. comm., 2012).

The first critical review of giraffe taxonomy was by Richard Lydekker (1904) in England. He examined type specimens of individuals that had been brought from all parts of Africa to Europe, noting especially distribution, coat pattern, coat colour and horn (ossicone) structure. Because of the very distinctive spotting pattern of reticulated giraffe from northeast Africa, he determined that this was a separate species, *Giraffa reticulata*. Giraffe from other parts of Africa he categorized as subspecies belonging to the species *Giraffa camelopardalis*. His classification has remained almost the same until the present, except that all nine kinds of giraffe are now described as subspecies belonging to the single species, *G. camelopardalis*. Russell Seymour (2001) briefly describes each of these subspecies in his PhD thesis.

It now is apparent that some or most of the races may actually be true species whose lineages have been separated from each other for up to a million and a half years (Brown *et al.*, 2007; Hassanin *et al.*, 2007). This may seem to some a matter of semantics, but it is far more than that. As we watch with horror the number of wild giraffe in Africa dwindle because of human population pressure, we must realize that it is not one species we are worried about, but perhaps eight or more. If any becomes extinct, as may soon happen to the Kordofan giraffe (*G. c. antiquorum*), it means the world has lost a distinct genotype that is gone forever.

The information in this chapter about numbers of giraffe and their distribution is based on current surveys by Julian Fennessy (2008, 2012a,b), co-founder and conservation scientist and trustee of the Giraffe Conservation Foundation (GCF) and Chair and founding member of the International Giraffe Working Group of the International Union of the Conservation of Nature (IUCN), who notes the impossibility of being completely accurate given the difficulties involved in collecting the data. Fennessy

(2012b) gives a chart of the recent status of giraffe in 98 reserves, national parks, ranches and unprotected areas of Africa.

Drastic decline of populations

In about 1960, in order to gather information on giraffe distribution, numbers and races (Dagg, 1962a,c), I simply wrote to the game departments or government agencies in charge of wildlife in each country that had giraffe among its wildlife; wildlife management was then in its infancy. I accepted gratefully the estimates of giraffe numbers and maps that were sent to me in airmail envelopes, but I had no way to check their accuracy. (Dagg and Foster (1976) contains this information on the distribution and status of giraffe in African countries along with updates in the 1982 edition.) At that time it was impossible to know how many giraffe there were in Africa, but there was no thought of them needing protection. In 1957, on the bus between Arusha and Nairobi, it was common to see giraffe strolling nonchalantly about near the road. Today there are still problems gathering accurate information about giraffe populations, especially with numbers decreasing rapidly because of human depredations – poaching, increasing agriculture, decreasing natural habitat, famine and war.

The decline in giraffe numbers recently is devastating. In the late 1990s there were about 140,000 giraffe in Africa (East, 1999), but during the next decade, with a huge 40% reduction the numbers fell to about 80,000 (Fennessy, 2012b). Using data collected by Julian Fennessy, the IUCN in 2008 declared the *peralta* race 'Endangered' and in 2010 the *rothschildi* race also 'Endangered' on the Red List of Threatened Species; for this latter race fewer than 670 individuals remained. The *reticulata* race and other races now under study may soon be considered subspecies of 'Concern' or 'Near threatened'; although *reticulata* has many more members than most races, its numbers are declining at a rapid rate. (Such designations depend on the rate of decline in numbers of a species or race; species with few animals are not considered at risk if they never were populous.) Ideally, perhaps all giraffe subspecies/races should be listed in a threatened IUCN category, a possibility being considered by the Giraffe Conservation Foundation (GCF).

Fennessy has investigated in depth the presumed past and present distribution of giraffe (Figure 11.1) and the status of the different races in Africa (Tables 11.1 and 11.2). Although we know from DNA samples that some of these races are disparate enough to be full species, they have not officially been declared such, so here we shall consider them as races even though they may (depending on whether the powers-that-be are 'splitters' or 'lumpers') soon become species. Further information about early naming of groups, locations of type specimens (both in the field and where the type animal is now stored) and descriptions of races (considered now sometimes suspect) are available (Dagg and Foster, 1982, pp. 47–48, 156–159). Descriptions of the coat pattern, coat colour and horns to define a race are downplayed here because they often vary within any one population and with the age of individuals (Dagg, 1962c). In some cases a race has

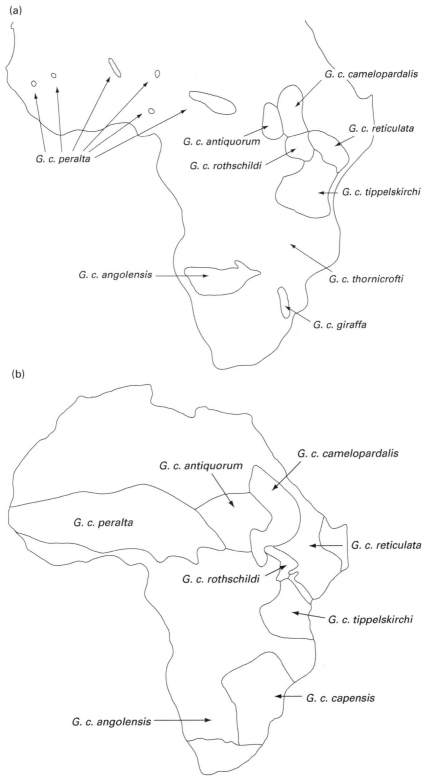

Figure 11.1 **a**, Subspecies of giraffe according to Dagg (1971), redrawn by Seymour (2001). **b**, Subspecies of giraffe according to Krumbiegel (1939), based on the work of Lydekker (1904) and published by Seymour (2001).

Table 11.1 An overview of giraffe *Giraffe camelopardalis* taxonomy, distribution and conservation status from Fennessy (2008).

Subspecies type	Common name	Distribution	Estimated numbers	Country status†	Subspecies status
G. c. angolensis	Angolan	Angola	–	Ex	Increasing
		Botswana	11,700	S/I	
		Namibia	6690	I	
		Zambia	Unknown	S/D	
G. c. antiquorum	Kordofan	Sudan	–	Ex	Possibly Extinct
G. c. camelopardalis	Nubian	Eritrea	–	Ex	Decreasing
		Ethiopia	160	D	
		Sudan	–	Ex	
G. c. giraffa	Cape/	Mozambique	–	Ex	Stable/
	Southern	South Africa	7880	S/I	Increasing
		Swaziland	80	S/I	
		Zimbabwe	5430	S/D	
G. c. peralta	Nigerian	Cameroon	1360	S/D	Decreasing
		Chad	800	D	
		CAR*	> 550	D	
		Mali	< 10	D	
		Niger	135	I	
		Nigeria	–	V	
G. c. reticulata	Reticulated	Ethiopia	> 140	S/D	Stable/
		Kenya	27,540	S/D	Decreasing
		Somalia	–	Ex	
G. c. rothschildi	Rothschild's	Kenya	> 140	D	Decreasing
		Sudan	–	Ex	
		Uganda	> 150	S/I	
G. c. thornicrofti	Thornicroft's	Zambia (Luangwa Valley)	1160	S	Stable
G. c. tippelskirchi	Masai	Kenya	17,330	D	Decreasing
		Rwanda	20	S/D	
		Tanzania	28,860	S/D	

† Ex, extinct; S, stable; I, increasing; D, decreasing; V, vulnerable;
* CAR, Central African Republic.

been described from the outward appearance of only one or two animals, or a giraffe was catalogued with a sole comment such as 'sent from Cape Town' or 'from Mombasa'.

The GCF and International Giraffe Working Group (IGWG) have been vital in obtaining information about the status of giraffe in many parts of Africa, publicizing this information widely, and helping to formulate and carry out plans to prevent populations of giraffe from becoming extinct. Other organizations also working to protect giraffe are listed at the end of this chapter. The current situation of each race and measures being taken to conserve it are included here. Subspecies are listed in approximate order of their danger of extinction. Note from Table 11.2 that reintroductions of giraffe are

Table 11.2 Taxonomy, distribution, abundance and status of the nine described giraffe subspecies. From Fennessy (2012b).

Subspecies	Common name	Distribution	1998 estimated numbers	1998 country status†	Subspecies status	2012 estimated numbers
G. c. angolensis	Angolan	Angola	–	Ex	Increasing	Extinct
		Botswana	11,700	D		6500
		Namibia	6690	I		10,000
		Zambia	Unknown	S		>50
G. c. antiquorum	Kordofan	Cameroon			Stable/Decreasing	Extinct
		Chad	Not reported	Not reported		162
		CAR*				75
		DRC**				Possibly *G. c. camelopardalis* or *G. c. rothschildi*
		Sudan				
G. c. camelopardalis	Nubian	Eritrea	–	Ex	Decreasing	Extinct
		Ethiopia	160	D		150
		Sudan	–	Ex		400 – possibly *G. c. antiquorum* or *G. c. rothschildi*
G. c. giraffa	Cape or Southern	Angola	–	S	Stable/Increasing	<50 – extralimital introduction
		Mozambique				<50 – reintroduced
		South Africa	7880	S/I		8000
		Senegal	Not reported	Not reported		50 – extralimital introduction
		Swaziland	>25	S/I		25
		Zimbabwe	5430	S/D		Unknown
G. c. peralta	Nigerian or West African	Cameroon	1360	S/D	Decreasing	See *G. c. antiquorum*
		Chad	800	D		See *G. c. antiquorum*
		CAR*	>550	D		See *G. c. antiquorum*
		Mali	<10	D		Extinct
		Niger	70	I		310
		Nigeria	–	V		Extinct

Table 11.2 (cont.)

Subspecies	Common name	Distribution	1998 estimated numbers	1998 country status[†]	Subspecies status	2012 estimated numbers
G. c. reticulata	Reticulated	Ethiopia	<140	S/D		<25
		Kenya	27,540	S/D	Stable/Decreasing	5000
		Somalia	–	Ex		Unknown
G. c. rothschildi	Rothschild's	Kenya	>140	D		350 – extralimital introduction
		Sudan		Ex	Stable/Decreasing	Possibly *G. c. camelopardalis* or *G. c. rothschildi*
		Uganda	>150	S/I		350
G. c. thornicrofti	Thornicroft's	Zambia	1160	S	Decreasing	750
G. c. tippelskirchi	Masai	Kenya	17,330	D		12,000
		Rwanda	20	S/D	Decreasing	<100 – extralimital introduction
		Tanzania	28,860	S/D		28,000

1998 estimated numbers: East (1999).

[†] Ex, extinct; S, stable; I, increasing; D, decreasing; V, vulnerable;

* CAR, Central African Republic; ** DRC, Democratic Republic of the Congo.

occurring in small numbers where giraffe have been wiped out and that some giraffe are being trucked to places where they never lived in the past for 'extralimital' introductions.

G. c. peralta West African or Nigerian giraffe

The West African giraffe was first described by Thomas (1898) from the skin, skull and cannon bone of a large female. The type specimen was shot near the junction of the Niger and Benue River in Nigeria, a place where they have long been extinct. This subspecies had an enormous range, spreading over what is now the western Sahara Desert, wherever there were permanent water sources, as early as the Palaeolithic. That they were present along the Mediterranean coast is confirmed by skeletons and rock drawings there; indeed giraffe lived in Morocco as recently as 600 AD (Ciofolo, 1995). At the end of the nineteenth century these giraffe were present in Gambia, Mauritania, Senegal, Mali, Nigeria and Niger, but during the next 100 years they have disappeared in every country but Niger because of extensive hunting, deforestation, development of agriculture, the construction of railroads and the installation of telephones (Ciofolo, 1995; Fennessy, 2008). In 1970 hundreds of giraffe lived in Niger, but by the mid 1990s few remained. Drastic measures were needed to save the giraffe, and fortunately these were put in place.

Niger is a dry country in the sahel region of Africa, one of the poorest in the world, but it has wonderful people, farmers and herders, because they are content to have giraffe roam freely among them. The vision for saving giraffe in Niger is not one of setting aside special reserves for them – after all, there are few fertile areas in the country – but of having giraffe live alongside the people of the various ethnic groups who become used to their presence and like to have them nearby. It is wonderful to see photos of giraffe and people going about their business almost side by side.

The giraffe world was devastated at the death in an August 2012 plane crash of Jean-Patrick Suraud, who was devoting his life to giraffe and who had written his PhD thesis on the *peralta* race in Niger (Fennessy, 2012c). Suraud (2011) had noted four reasons why these giraffe are unique:

- they are the only remnants of the *peralta* subspecies, sometimes called 'white giraffe' because of their pale colouring, which once spread across most of north Africa;
- they live among dense populations of people;
- there is no special park or reserve set aside for them; and
- they have no natural predators (aside from people).

Suraud determined the distribution of these giraffe at both population and herd levels through direct observations, and measured habitat selection of individuals by using GPS satellite collars – he kept a permanent identification file for each animal. Their distribution, which changed seasonally, was correlated with food availability (Le Pendu and Ciofolo, 1999). During the dry season the giraffe avoided villages, but did approach them at night where food resources, both quantity and quality, were higher (tree density, granaries). The newborn giraffe were mostly male, so Suraud (2008) suggested there might be a bias investment toward males indicating better than average life conditions.

Figure 11.2   Photo of Jean-Patrick Suraud, giraffe devotee, killed in a plane crash in August 2012. Photo by Andy Tutchings, GCF.

(We saw in Chapter 4 that where hunting for bush meat was high, there was a birth bias toward female calves.)

Giraffe in Niger were almost extinct in 1996, with only about 50 animals remaining, but they have been brought back from the abyss thanks to dedicated internationally funded projects and scientists such as Jean-Patrick Suraud, Isabelle Ciofolo and others working locally with the Association to Safeguard the Giraffes of Niger (ASGN) and the government. From 1996 to 1999, the annual growth rate was a spectacular 19% (Suraud *et al.*, 2012) following the cessation of a period of severe poaching. In 2012 there were estimated to be 310 giraffe, but the race remains classified as 'Endangered' by the IUCN Red List assessment criteria; there should be at least 400 adult giraffe to ensure a viable breeding *peralta* population (Suraud, 2008). The challenge is how to have giraffe prosper when they survive among farmers and nomads who themselves live in poverty.

Related to this aim have been studies on Population and Habitat Viability Assessment (PHVA) processes. For example, using simulation computer modelling for the Niger giraffe, giraffe people in workshop sessions considered such possibilities as the situation remaining as it is, habitat degradation, and the possibility of establishing a new population elsewhere, along with possible ecological events, political events and disease outbreaks (Desbiez and Leus, 2009a,b). Recommendations for the preservation of giraffe in Niger include:

- developing monitoring systems for early detection of catastrophes along with plans on how to react to them;
- maintaining the current giraffe population in the 840 km² 'giraffe zone' at all costs;
- ensuring that giraffe can safely expand outside their current range, should this occur;
- establishing another giraffe population as extra insurance against some catastrophe, taking great care with reintroductions. This new region should have a carrying capacity for at least 500 giraffe;
- monitoring continually the rate of any habitat loss so these data can be added to the models; and
- continuing to refine the model of the existing population, adding estimates of age and sex.

American Jennifer Margulis (2008) had been thrilled to see people and giraffe intermingling in Niger in 1992–1993 when she had worked there for a non-profit development agency and become a friend of Isabelle Ciofolo. When she returned for a visit in 2007, she was heartened to find that the number of giraffe was increasing: in 2007, 164 had been photographed for the census by ASGN, up from 144 in 2006, and in 2008 an amazing 21 or more calves were born.

Margulis reports that earlier the government had outlawed hunting of giraffe (although several have been killed by bush taxis, buses and poachers). The cutting down of trees was to be banned and $40,000 was to be spent annually on anti-poaching, but these mandates had not been carried out. In 1996 after a *coup d'état*, the new leader had decided to give two giraffe each to the presidents of Burkino Faso and of Nigeria. When the forestry office refused to go along with this, he sent in the army. More than 20 giraffe, nearly a third of the total population, was wiped out. The next president of Niger who took power in 1999 wanted to give a pair of giraffe to Togo's president. With the aid of the Togolese Army, helped by local villagers and the forestry service, two giraffe were finally captured after a pursuit of 3 days. One died on the way to Togo and the other soon after. Currently illegal hunting in Niger results in a few deaths of giraffe each year, including some animals that cross into Nigeria.

French researcher Isabelle Ciofolo, who was the first person to study giraffe in Niger and has devoted many years of her life to them, helped found the ASGN in 1994. Its mandate is to protect giraffe habitat, educate the local people about this animal, take an annual census of giraffe and provide microloans and other aid to villagers who live in the giraffe zone. Many wells have been constructed with ASGN funds which benefit a whole village. One local Zarma woman bought sheep and goats with an ASGN loan which she fattened up and sold for a profit. 'Giraffes have brought happiness here. Their presence brings us lots of things', she said. ASGN efforts are supported by funds from environmental groups around the world such as the African Wildlife Foundation and zoos in France, UK and USA. The *peralta* giraffe are relatively safe if they remain in Niger, but as the human population grows there, giraffe herds will be increasingly stressed. People seeking wood are destroying the 'tiger bush' habitat of giraffe during the rainy season (Suraud, 2006), despite laws against this, so the giraffe must extend their range into new areas where there is sufficient food although here, too, there will be friction with people.

The problem of giraffe–human conflict will only increase in the future. Farmers continue to clear areas of native vegetation so they can grow crops and keep livestock, while giraffe sometimes feed upon their cowpeas and mangoes. To assess the situation, Romain Leroy and three colleagues (2009) carried out a survey of chiefs and farmers (who on average support 15 people each) in the giraffe zone, asking about their perception of this animal. Of the 227 farmers who grew cowpeas, 63% had suffered damage from giraffe in their cowpea fields at harvest time, and 32% in cowpea granaries which were nearer their houses. Of the 113 who grew mangoes, 82% had suffered losses from giraffe. Most respondents thought of giraffe as a disadvantage, but some liked them for their economic contribution through tourism or culture. However, tourism did not benefit most families, and was thought to be distributed inequitably among the villages.

Although most farmers did not want giraffe eating their crops, they did little to prevent this such as organizing harvest surveillance teams, although some had created barriers around their granaries. Most mango growers had built fences around their orchards, but they admitted these were more to keep out stray dogs and children than giraffe. Indeed, although the perception of farmers about giraffe was mostly negative, there was no evidence of this negativity in daily life. Giraffe were amazingly well tolerated, with crop protection measures minimal and poaching rare. We must be grateful to the farmers for their forebearance and hope that it continues.

In 2012, a management plan was approved for the giraffe zone of Niger between the Kouré Plateau and Dallol Bosso (Salifou, 2012). Recommendations are to:

- reinforce and support community projects related to the giraffe;
- work toward management and restoration of the giraffe zone;
- support the development of activities to generate money for women;
- construct solar panels to provide energy and reduce wood cutting in the zone; and
- arrange a corridor 20-km long for giraffe to obtain drinking water.

The current population of giraffe in Niger seems close to saturation with respect to the present habitat. If it is to reach a sustainable level, giraffe must expand into new areas; observations that individual animals have travelled for long distances in Niger may, in fact, be related to this possibility (Suraud, 2009). However, it is essential to have a management plan to ensure the West African giraffe does not become extinct.

G. c. antiquorum Kordofan giraffe

The type giraffe for this race was first described by Jardine in 1835, with the type location given by the vague appellation 'Sennar and Darfour' in the Sudan. A second claim is for a male shot at Baggar el Homer, Kordofan. The race occupied the west and southwest corner of present-day South Sudan and into adjacent Democratic Republic of the Congo. It is assumed to be synonymous with Lydekker's *congoensis*.[1]

The recent decline in numbers of Kordofan giraffe in Democratic Republic of the Congo has been steep, from 345 giraffe in 1993 to 62 in 2002 and fewer than 80 in 2012 (Amube *et al*., 2009; Marais *et al*., 2012d); all reside in Garamba National Park or its three adjacent hunting reserves. There is huge concern for their conservation, especially recently with South Sudan military crossing the border to poach for bush meat. There had been a cultural taboo against poaching because giraffe meat was believed to cause leprosy, but this superstition has largely died out. Security had been so lax that the Lord's Resistance Army (LRA) was able to set up base camps within Garamba National Park, one lasting from 2005 to 2009. At present, protection of the giraffe is stabilized with the assistance of the Congo military (FARDC) and UN Special Forces (MONUSCO) (Marais *et al*., 2012d). Such forces must remain in place if the giraffe are to be saved.

Marina Mònica and Nuria Ortega (2012) wanted to discover why there were so many fewer giraffe in Garamba National Park, independent of poaching. Could it be limited forage? Or high predation? To find out, they have recently collared giraffe so that their future movements and activities can be traced. The animals were too skittish to be approached on foot, so darting was done by light aircraft after the plane had herded each individual to an open area easily accessible by vehicle. After a successful darting, the ground crew had 3 min to reach each giraffe and assist its fall by holding its head and neck erect. A veterinarian then administered an antidote which left the animal awake, but held down and kept calm by a covering over its eyes and ears. While in this state the team quietly placed the collar around its lower neck, took blood, skin and hair samples, and checked its vital signs. When this was done, the group backed off so the giraffe could get up and escape. This research is ongoing.

The Central African Republic (CAR) now has fewer than 170 giraffe so their status is also dire (Fennessy, 2012b). Philippe Bouché and colleagues (2009) reported armed conflict and extensive poaching for bush meat often by men from distant places using automatic weapons. If these issues are not addressed and CAR's borders not made secure, then giraffe numbers and biodiversity in general will continue to decline. Wildlife recovery is only possible if a strong regional political commitment is expressed and implemented in the field. The presence of the national army must be permanent in northern CAR to establish and maintain discipline and security. Indeed, measures should be international, since the same problems plague not only CAR, but also its neighbours Chad, Democratic Republic of the Congo and South Sudan. The authors recommend that these four countries band together to lobby international donors so that programmes can be carried out to save the wildlife in the huge shared area where few people live.

The most recent news from this area is devastating (Groo, December 2012). Poaching for bush meat of elephant, rhinos, zebras and giraffe has reached high levels from hunters entering CAR from South Sudan; even security forces are involved in trafficking. Many

---

[1] Giraffe in Garamba National Park are called *G. c. congoensis* on the Park's website, perhaps a chauvinist conceit because this subspecific appellation is no longer accepted by taxonomists.

of the poachers have been caught red-handed, but a lack of laws and weak institutions prevent them from being penalized.

In Chad, Waza National Park contained about 600 giraffe in 2009, most of them primarily in the *Acacia seyal* zone where one of their favourite food trees flourishes (Omondi *et al.*, 2009). This was a significant decline from 20 years earlier, attributed to increased pressure of livestock and illegal hunting. That same year 612 giraffe were counted in Zakouma National Park (Potgieter, 2009).

Mitochondrial studies of DNA variability indicate that *antiquorum* includes all giraffe populations from Cameroon (where populations are good although poaching continues), Chad, CAR and perhaps South Sudan (Hassanin *et al.*, 2007) where numbers are rapidly decreasing; these races do not, therefore, belong to *peralta* but are more closely related to the *rothschildi* and *reticulata* subspecies.

G. c. rothschildi Rothschild's, Ugandan or Baringo giraffe

Kenya is passionate in its conservation measures for the three races living in its boundaries: *rothschildi*, *reticulata* and *tippelskirchi*. Most other nations now have only one race. Kenya has the earliest known giraffe fossils, suggesting that it may be the epicentre of the species' radiation across Africa (Brown *et al.*, 2007). The Kenya Wildlife Service, with the help of local and international experts, will soon have a national strategy for giraffe conservation, the second ever for an African country after Niger. It will provide guidance for the management of each of its three races and publicize their decreasing numbers (Muller, 2010b).

The Rothschild's race was first described by Sir Harry Johnston, a Colonial Administrator of Uganda, who encountered the giraffe in Kenya in 1901 and sent the skin and skulls of two males and two females back to the British Museum. He emphasized the males as distinctive with five 'horns', although five 'horns' occur in other races too. The type animal was shot near Lake Baringo by Major Cotton. The race was named for Lionel Walter, 2nd Baron Rothschild, founder of the Tring Museum in England, who was the eminent zoologist to first describe it (Thomas, 1901). It subsumes the name *G. c. cottoni*.

These giraffe were once present in current-day South Sudan, northeast Kenya, Uganda and perhaps Ethiopia, but their range now is so much reduced by agricultural development, human settlement and poaching that there is fear of their extinction. *G. c. rothschildi* which now includes only about 670 animals was declared an Endangered subspecies by IUCN in 2010 (Fennessy and Brenneman, 2010).

The small remaining populations are scattered, which unfortunately prevents them from interbreeding. Rothschild's giraffe in Uganda are present in Murchison Falls National Park (down from 2500 in the 1960s), the only remaining population in its natural range which has not been reintroduced or supplemented with giraffe from other populations. Other Rothschild's occur in the Kidepo Valley National Park in Uganda (fewer than 20 in all Uganda in 2012), and in Kenya where it has been reintroduced to various sites: Lake Nakuru National Park (see also Chapter 3), Ruma National Park, Soysambu Conservancy, Mwea Natural Reserve and Giraffe Manor (Brenneman *et al.*, 2009a). (Giraffe Manor/Hotel in Nairobi which now features almost a dozen giraffe has

Figure 11.3 Four Rothschild giraffe being ferried across Lake Baringo to reach their new home from which their race had long been exterminated. Photo by Zoe Muller.

accommodation for guests to sleep there; part of their payment supports the African Fund for Endangered Wildlife (Anon., 2012d).) Six extralimital introductions have also recently taken place in Kenya (IUCN, 2011) which now has a significant number of giraffe of this race, although none in their native range with the exception of eight shipped to the Lake Baringo area.

Zoe Muller (2011) of the Giraffe Conservation Foundation and Coordinator for the Rothschild's Giraffe Project describes this epic undertaking in February 2011, 4 years in the planning, to translocate giraffe across Lake Baringo so that this race could be returned to the native habitat from which it had been driven or killed over 70 years earlier; it is sometimes called the Baringo giraffe. Eight healthy animals from Soysambu Conservancy where the Rothschild's Giraffe Project is based were trucked 160 km north to the shores of Lake Baringo, herded carefully in two groups of four onto a barge with canoes fastened to either side for stability and ferried for an hour to the opposite shore. The giraffe had stood during the long night drive, in the holding pen, and on the barge, so when they finally reached their new home, the Ruko Conservancy, they galloped up the strung-together chute to freedom. This community conservation project, the first reintroduction of Rothschild's giraffe to its native range in Kenya, has no access by road, so it should be relatively easy to keep the animals safe from poachers. It is hoped that with such wildlife introductions local people can find jobs through wildlife-based tourism, guiding, tracking and working at resorts.

Muller (2012b) reports that these translocated giraffe have settled down nicely in their new home, with two females having given birth. She (2010c, 2012a) continues to work for the conservation of this subspecies.

More recently, three giraffe of this endangered race were translocated 60 km by truck from Kigio Wildlife Conservancy to Solai Conservancy near Nakuru, Kenya, thus extending their range (Sarginson and Nissen, 2012). However, translocation is difficult, and two giraffe died during the undertaking.

Recently Colin Groves and Peter Grubb (2011) have proposed subsuming the Rothschild's race into that of the Nubian giraffe, creating eight species (not subspecies) for the giraffe.

Case study

The overland translocation in 1983 of 27 *rothschildi* giraffe from Kenya's Rift Valley to Ruma National Park in Kenya helped increase the number of giraffe there to 69 in 1999 (Awange *et al.*, 2004). In 2002, an expedition of five scientists visited the Park to see how these giraffe were doing. They counted about 75 in all, and were told that in the past 19 years one giraffe had died of natural causes and seven had been killed at one time by lightning (really also a natural cause). They noted that the giraffe fitted well into the environment because they did not compete with other species for food, although they did share the ample water supplies. They appeared free of disease although often with hoof abnormalities.

However, there were problems:

(a) browse favoured by giraffe such as *Acacia drepanolobium* had decreased in the Park;
(b) giraffe sometimes encroached on crops outside the Park, destroying sorghum and sunflower, while people sometimes entered the Park where there was no fencing to destroy wildlife;
(c) people also let cattle range within the Park borders and cut down trees, creating soil erosion;
(d) there were hard feelings between the Park warden and the farmers, who sometimes started uncontrolled fires to express their dissatisfaction with the Park's administration; and
(e) few wildlife personnel and only one vehicle were available to patrol the giraffe's range.

The authors suggest ways in which the situation at Ruma could be improved:

(a) involve schools so that local residents would become aware of the benefits of land use management;
(b) hire local people for increased wildlife management positions;
(c) meet with local chiefs to plan collaborative coordination between their people and the Park;
(d) develop and distribute educational materials about the Park;
(e) make sure that fencing is completed and not allowed to deteriorate; and
(f) pave the roads leading into the Park which became impassable during rainy seasons.

G. c. camelopardalis Nubian giraffe

The Nubian giraffe was the first race to be exhibited alive in Europe because it could readily be captured by Arabs, sent by boat down the Nile to Cairo and then on north into Europe. Linnaeus' description in 1778 of the subspecies came from a second-hand description of a giraffe captive in Cairo. He had not seen the animal, so first named it *Cervus camelopardalis* from 'Sennar and Aethiopia', thinking that it must be related to the European red deer because its horns were covered with skin as were the antlers of deer during their growth. In 1788 the type location was revised to 'Sennar between Upper Egypt and Ethiopia'. The 400 or so giraffe left in South Sudan may actually belong to the *antiquorum* or *rothschildi* race (Fennessy, 2012a).

Information from the Giraffe Conservation Foundation (Marais *et al.*, 2012b) in Ethiopia indicates that there are now fewer than 150 Nubian giraffe in the Gambella National Park of Ethiopia, fewer than 20 giraffe in Omo National Park which may in fact be Rothschild's race rather than Nubian and that some reticulated giraffe may possibly survive in the Ogaden region bordering Somalia.

Data for South Sudan indicate that in 2007 there were about 400 giraffe in Boma National Park and some also in Bandingilo Reserve (Marais *et al.*, 2012a). However, at that time no giraffe were sighted in aerial surveys over Jonglei area or Southern National Park. There is uncertainty about the safety of giraffe in this country given military interventions, and whether these giraffe do indeed belong to the Nubian race.

G. c. thornicrofti Thornicroft's giraffe

This race, confined to Zambia, was first described by Lydekker (1911) from a giraffe shot by Mr Thornicroft. There were several populations of giraffe in the area of the Luangwa Valley which were protected from then on. In 1974 there were estimated to be about 300 animals (Ansell, pers. comm., 1974), in 2008 about 1160 and in 2012 between 750 and 1000 (Fennessy, 2008, 2012b); the variation in numbers probably does not represent large changes in the populations, but rather poor guesstimates.

These numbers for this distinct group of giraffe are small, but the animals themselves are significant because one person, Philip Berry, has been willing and able to follow the destiny of some of them for over 50 years. The importance of such information gathered over decades is incalculable. In 1973 and again in 1978, he published monographs of general information on their activities. Then recently, he has teamed up with Fred Bercovitch (2009a,b, 2012) to present invaluable long-term data valid for wild *thornicrofti* giraffe including the longevity of many individuals (one female lived for 24 years), calving rates and timing for females (one 20-year-old female never bore a calf), herds sizes and compositions over many years, and long-term friendship between individuals (often between mothers and their daughters).

Until recently there was uncertainty about whether these giraffe constitute a distinct subspecies or even species. Their home range is 400 km south of *tippelskirchi* giraffe in Tanzania, and somewhat further away from other giraffe in western Zambia, Botswana and Zimbabwe. In addition, their coat pattern tends to be diagnostic, with smaller spots on the hindquarters than on the forequarters (Seymour, 2001). Are the *thornicrofti* animals a true subspecies? Very recent genetic research shows that they are indeed a distinct subspecies, genetically isolated from all other races (Fennessy *et al.*, 2013).

G. c. reticulata Reticulated, netted or Somali giraffe

This race was first described by de Winton (1899), who chose the type location as the Loroghi Mountains in Kenya. Its large range initially extended from Lake Turkana, Kenya, in the west, the mountains of Ethiopia in the north, the east coast of Africa (which is actually too dry for giraffe) and the Tana River, Kenya, in the south. Various racial names are subsumed into this race: *nigrescens*, *hagenbecki* and *australis*.

Reticulated giraffe, 450 of them, are frequent in zoos and probably, therefore, the most commonly known race, striking in their rich brown polygonal spots separated by thin white lines. Reticulated giraffe are the most distinctive race (as they were claimed to be over 100 years ago when they were declared a distinct species) with a DNA indicating virtually no interbreeding with other races of giraffe for over 1.5 million years (Brown *et al.*, 2007). However, individuals have been reported in east central Kenya who have spotting patterns intermediate between those of reticulated and Masai giraffe (Stott and Selsor, 1981), and which have hybrid genetics of these two races (Brenneman, 2004).

Reticulated giraffe are becoming rare in the wild in northeastern Kenya and especially, it is feared, in Somalia and southern Ethiopia. In the past decade, numbers of this race have fallen from 28,000 (data from IUCN African Antelope Database) to fewer than 5000, a decline of more than 80%. Outside protected areas, the remaining range of reticulated giraffe is characterized by growing populations of people living in poverty, refugees with automatic weapons from local and regional conflicts, overstretched security forces and environmental degradation (Doherty *et al.*, 2012). This catastrophe is blamed largely on poaching for what is called 'bandit food'. Local people make trophies from their tail hair, bone marrow and hides; long buckets fashioned from their neck skin are sometimes used as buckets to pull water from wells.

Working for the conservation of reticulated giraffe is PhD student John Doherty (2011) of Queen's University in Belfast, Ireland, who, with his team, is based in the Samburu National Reserve, Kenya. These giraffe can be seen in Samburu, Buffalo Springs and Shaba National Reserves, Ol Pejeta Conservancy and Lewa Wildlife Conservancy as well as other private and communal protected areas in northern Kenya. They also inhabit extensive arid rangelands in northeastern Kenya where they are not protected and where drought occasionally occurs.

Doherty's research projects which should help preserve giraffe populations in the future include the following.

- Behavioural ecology, working with at least 500 giraffe who can now be recognized individually, to understand their basic needs.
- Bioacoustics, exploring the implication of the now-known ability of giraffe to hear sounds too low for human ears. This was an exciting discovery, implying that giraffe

are more social than previously thought. Individuals may not be together physically, but they can probably communicate and apparently are members of a larger community of which humans are unaware. (See Chapter 4.)

- Remote sensing so that by affixing telemetric devices to free-living reticulated giraffe, researchers can measure their movements, home ranges, behaviours and energy expenditures. (A number of giraffe in Namibia, South Africa, Botswana and DRC now wear well-tested head harnesses that provide such information.)
- Population dynamics which involve collecting information from people living in northern Kenya and surrounding areas about the distribution and numbers, both present and past, of reticulated giraffe.

Doherty has initiated a trial idea of having local people name a specific giraffe after one of their relatives or friends, so that in the future they will have a proprietary interest in that animal and not want it harmed. People did recognize 'their' giraffe when they encountered it in the wild, and seemed pleased to have a connection with it. This concept could be important in the future. Also connected with conservation efforts, Francois Deacon (2012) working in South Africa named the eight giraffe whom he collared after the sponsor who paid for each GPS collar as a token thank you.

Case study

A new initiative for reticulated giraffe has been organized in Garissa County of northeastern Kenya where many Somali pastoralists live who moved here after the collapse of the Republic of Somalia in 1991 (Hussein, 2009). Giraffe were heavily poached at first, but some found refuge near Garissa town (where there was browse, cover and security) and among the riverine vegetation of the Tana River. These giraffe, about 30 of them, some injured with gunshot wounds, became the founders of the 125 km^2 Garissa or Bour-Algi Sanctuary. However, their habitat is currently endangered by subsistence farming, extensive grazing of cattle, and such practices as settlement encroachments, infrastructure development, expanding farmlands, charcoal burning, wood cutting, sand harvesting and poaching. A new sewage project also threatens the giraffe in the sanctuary (Ombuor, 2011). They are in danger of extinction unless something is done.

The solution is hoped to be the Garissa Giraffe Conservation Program, which is a community-based initiative, supported by the Giraffe Conservation Foundation, National Museums of Kenya and WildlifeDirect (Hussein, 2009). Its aims are to empower pastoral Somalis to:

- monitor the giraffe in the Bour-Algi Giraffe Sanctuary;
- monitor all threats to the animals;
- educate local and other people about giraffe and the environment;
- make management recommendations for the permanent compatibility of wildlife, livestock and pastoralists in the area; and
- encourage tourism and other sources of revenue to finance these initiatives.

Community-based programmes such as this are crucial for the survival of small wildlife populations such as that of giraffe in the Garissa Sanctuary. It is the people who live near them who realize the dangers giraffe face and know how best to protect them. The small Bour-Algi Giraffe Sanctuary cannot expect much help from centralized or distant governments, but can take advantage of the extensive expertise of the three outside groups which support it.

G.c. giraffa Cape or southern giraffe

This race was encountered not by Europeans coming from the north, but by Boers exploring from the south. The type specimen was shot in 1761 near Warmbad in Namibia, just north of the Orange River (Brink, 1954). Originally the race had an enormous range from Namibia, southern Botswana, eastern low veld of South Africa, western Zimbabwe to southwest Mozambique, but now it is much restricted because of human population pressure. They no longer occur in Namibia. This race was never found as far south as the Indian Ocean, presumably because the weather there is too cold in winter. In the South Africa low veld in 1957, a large number of giraffe already weakened by a low-quality diet died when the temperature approached freezing coupled with wind and rain (A. Matthew, pers. comm., 1958). Similar conditions caused death to many giraffe and other animals in 1968 in then-Rhodesia (Jubb, 1970).

   G. c. giraffa includes the former races *G. c. capensis* and *G. c. wardi*. Giraffe of this race have been reintroduced to Hluhluwe-Umfolozi National Park in KwaZulu-Natal and to small game reserves throughout South Africa including extralimital translocations in the south, to attract tourists in areas where giraffe had not been known before from colonial reports or rock art. However, giraffe bones have recently been discovered from Middle Stone Age deposits within KwaZulu-Natal (Cramer and Mazel, 2007). Giraffe seem to have been in this province 1000 years BP (the date of the most recent excavations with giraffe remains) but had died out or been extirpated by 220 BP (the first date of written accounts). Their demise could have been caused by disease, drought or hunting by people. This new information means that giraffe are not 'aliens' to the area as had been presumed, an important distinction because South Africa's National Environmental Management: Biodiversity Act (Act No. 10 of 2004) prohibits the introduction of an alien species into a habitat where it does not occur naturally. As an alien, giraffe should be removed from KwaZulu-Natal to the distress of visiting tourists; as a non-alien, managers can wholeheartedly rejoice at their presence. In 2012 there were about 8000 of these giraffe in South Africa, 25 (extralimital) in Swaziland, an unknown number in Zimbabwe, some (extralimital) in Zambia, a few (extralimital) in Angola and over 50 reintroduced into Mozambique and Senegal (Fennessy, pers. comm., 2012).

G. c. angolensis Angolan giraffe

Angolan giraffe are bucking the downward trend of this species across Africa (Tutchings, 2011). Their numbers are actually increasing, although despite its name it no longer occurs in Angola.

Captain Hendrik Hop ventured north of the Orange River into Namibia (then South West Africa) in 1761 where he encountered giraffe for the first time. One was captured and killed near Warmbad to become the type specimen of this race (or could that specimen have been a Cape giraffe?); this was the first known record of a professional hunt of any wildlife species in Namibia (Fennessy, 2008). Shortridge (1934) much later, but still while Namibia was relatively untouched by European civilization, reported that about 400 giraffe ranged throughout South West Africa. He estimated that about half of these were in the far northeast Kunene region of the northern Namib Desert where there is minimal food and water resources, making it an incredibly harsh environment. The rest of them occurred in the Grootfontein and Caprivi areas. Since that time, the number of giraffe in the Kunene area has increased to about 900 animals, with a number of ups and downs during this period (Fennessy, 2008).

Chris Brown (2010) has researched the historic distribution of giraffe in the Greater Fish River Canyon Complex of southwest Namibia. Giraffe and other large wildlife species were hunted to extinction there by 1840 to make room for small-stock farming, although this was a totally unsuitable endeavour given the area's extreme aridity and highly variable rainfalls. In the 1990s, the area was deemed more suitable for wildlife and tourism than farming, so Brown recommends that giraffe be reintroduced where they had once flourished:

(a) at least 20 at a time be brought into suitable areas (because small token introductions invariably fail);
(b) all introduced wildlife be from other similar dry environments such as Namib or Karoo;
(c) no alien plants or animals be introduced;
(d) human neighbours be agreeable to the introductions; and
(e) annual censuses be carried out to allow ongoing, up-to-date management.

Julian Fennessy (2003, 2006; Fennessy et al., 2003) was the first to study the desert-dwelling angolensis giraffe in three areas of northwestern Namibia where the densities are as low as anywhere in Africa at 0.01 giraffe/km^2. In 2008 he published a thorough account of the status of G. c. angolensis (earlier called G. c. infumata) in Namibia where the giraffe numbered about 4000 in the 1970s and 10,000 in 2012 (Fennessy, 2008, 2012b). All the giraffe in this country belong to the angolensis race, including those in the Etosha National Park and the Caprivi strip (Brenneman et al., 2009b). Angolan giraffe in the Namib Desert inhabit riparian woodlands of the six most northerly rivers and their tributaries (Fennessy, 2009). Because of their desert habitat giraffe populations are sparse, forage density is reduced and males have to walk a long way to find females in oestrus with whom to mate. Giraffe are classified by the Namibian government as 'specially protected', yet they can be legally hunted with a specific hunting permit. ('Specially protected' species in other countries such as Kenya, Niger, Swaziland, Tanzania and Zambia may not be hunted or killed.) Fewer than 50 Angolan giraffe live in Zambia, and giraffe in the nearby Hwange National Park of Zimbabwe may also belong to this race (Chamaillé-Jammes et al., 2009).

Figure 11.4   Julian Fennessy and his team discuss their research plans for the Angolan giraffe in Namibia. Photo by Julian Fennessy.

Angolan giraffe are also present in Botswana (pop. ~6500 although there is reportedly heavy poaching) (see Chase, 2011, for information about northern Botswana). They occur in the Kgalagadi Transfrontier Park, which comprises two earlier adjacent parks in South Africa and Botswana, respectively, Kalahari Gemsbok National Park and Gemsbok National Park. In this Park, inaugurated in 2000 and located largely within the southern Kalahari Desert, Francois Deacon (2012) is carrying out research including the use of GPS collars on eight giraffe.

Because of a decline of giraffe numbers, Kylie McQualter (2012) is organizing ecological research in the Chobe River front areas and in the Okavango Delta. Aerial surveys have been carried out across northern Botswana to determine giraffe numbers and distribution, and four GPS head harnesses have been fitted on females to document their daily and long-term movements. Information on population dynamics, herd composition and diet is also being collected.

This enterprise in Botswana is under the umbrella of the recently established Large Herbivore Ecology Programme of Elephants without Borders (EWB) (Tutchings, 2012). The initiative should make conservation efforts of wildlife in Africa more efficient by:

- pooling of resources, expertise and technical know-how; and
- sharing of information to prevent wasteful repetitive efforts.

Along with this much-needed endeavour, the president of Botswana has announced that commercial hunting of giraffe will be indefinitely suspended in 2014 (Anon., 2012f) (although limited traditional hunting will be allowed for some local areas which have a designated wildlife management plan). Hunting was a seasonal and necessarily limited activity (there are only so many giraffe to kill), whereas photographic tourism (which will now be encouraged) occurs all year around, so this is an important step forward for the economy.

Because *G. c. angolensis* has significant genetic differences from other subspecies as well as being isolated reproductively, its population (like others) constitutes a unique evolutionary unit that may be approaching species' level status. As another matter of taxonomic interest, giraffe in desert areas of Namibia have a pale colouration and so might be referred to as a separate 'eco-type' or 'biotype', but they are not genetically distinct from the other Namibian giraffe (Fennessy, 2008).

For seven giraffe from the Etosha National Park, Ken Carter and colleagues (2012) have developed 11 microsatellite markers using 454 sequencing, information which will be invaluable for conservation, population and quantitative genetic studies.

G. c. tippelskirchi Masai or Kilimanjaro giraffe

In 1898 Paul Matschie, a German, described two giraffe types from southern East Africa as *G. tippelskirchi* from Lake Eyassi, Tanganyika (now Tanzania), and *G. schillingsi* from Taveta, Kenya. Lydekker (1904) combined these two groups into a single race, *tippelskirchi*, whose range extends north through the Serengeti Plains and Masailand in Tanzania and Kenya, to the African coast in the east, south to the Rufiji River, and west to Lake Rukwa and Lake Tanganyika. Within Kenya they inhabit the Masai Mara, north to

Figure 11.5   Collared *G. c. peralta* giraffe with spotting much like that of several other species. Photo by Jean-Patrick Suraud.

Lake Naivasha and east to Tsavo East and West, plus extralimital introduction of giraffe into Shimba Hills National Park south of Mombasa. Giraffe still roam freely in and around Nairobi National Park, but human populations and fences will soon limit their future movements.

There are 35,00–40,000 Masai giraffe in Kenya and Tanzania which makes this the most populous race, but it too has suffered drastic losses recently of up to 60% (Fennessy, 2012b). They are being illegally killed for bush meat, especially in the Serengeti National Park and Selous Game Reserve in Tanzania and the Amboseli National Park in Kenya, and suffer also from illegal livestock disturbance and declining woodland cover. In the areas surrounding the Masai Mara Nature Reserve in Kenya, giraffe are being pushed out by growing human populations and the conversion of native habitat to commercial crops. The Serengeti Giraffe Project, headed by Megan Strauss and Craig Packer (2010), was initiated in 2008 to study the status and ecology of the Masai giraffe in Tanzania's Serengeti National Park, whose numbers had been decreasing precipitously.

Giraffe never occurred naturally in Rwanda, but two male and four female Masai giraffe were introduced into the southern part of Akagera National Park in 1986. The current number in the Park is now about 100 individuals (Marais *et al.*, 2012c).

Further research is needed

Before the era of DNA, it was assumed that there were nine races of giraffe. Now, after extensive genetic analysis, it seems possible that some of these are species rather than races judging from their DNA and their reproductive isolation from each other. The problem will need taxonomic consideration by both splitters and lumpers. Russell Seymour (2001) supported at least six valid subspecies using both morphological and genetic analyses. Based on their genetic research, David Brown and his team (2007) reported a minimum of six, and Alexandre Hassanin and colleagues eight distinct genetic groups that were not interbreeding. Colin Groves and Peter Grubb (2011) proposed eight full species of giraffe (with the Rothschild's race subsumed into the Nubian race), although not all populations have had their DNA sampled and analysed, and their results are based on published studies rather than primary research. The Rothschild's and West African giraffe are now included on the IUCN Red List as endangered subspecies, and it is probable that other races will soon have elevated conservation ranking as well to help ensure their future. Or possibly giraffe of all races will be listed at least in the 'Threatened' category of the IUCN. Taxonomic research should continue so that the race of all giraffe populations can be identified (Brown and Brenneman, 2006). This information will relate to their conservation management. Giraffe populations for which we need more genetic information live in the following areas (Fennessy, 2012b):

1. Cameroon – northern areas including Waza, Bouba Ndjida, Faro and Bénoué National Parks and surrounding hunting zones;
2. Central African Republic – Manovo-Gounda-St Floris National Park;
3. Chad – Zakouma National Park;
4. Democratic Republic of the Congo – Garamba National Park;
5. Ethiopia – Gambella National Park;
6. South Sudan – Boma National Park;
7. Southern Tanzania – southern areas including Katavi, Ruaha and Selous National Parks and surrounding areas;
8. Zimbabwe – Hwange National Park and surrounding areas; and
9. Zambia – Sioma-Ngwezi National Park and surrounding areas.

Organizations involved with giraffe conservation

- ASGN Association to Safeguard the Giraffes of Niger.
- GCF Giraffe Conservation Foundation – contact Julian Fennessy PhD, http://www.giraffeconservation.org.
- GOSG Giraffe and Okapi Specialist Group – now initiated under IUCN.
- IAGCP International Association of Giraffe Care Professionals; this important association was founded in 2010 by Amy Phelps of the Oakland Zoo, Paige McNickle of the Phoenix Zoo and Lanny Brown of the Phoenix Zoo. It meets every two years.

- IGWG International Giraffe Working Group of the IUCN – contact Julian Fennessy PhD.
- IUCN International Union for the Conservation of Nature.
- SSC Species Survival Commission, a commission of the IUCN.

Information about these groups is available on the Internet.

Appendix
Parasites and pathogens

Organisms that have been found in or on giraffe:

Ticks

Amblyomma cohaerens	Kenya	Walker, 1974
Amblyomma eburneum	Ethiopia	Theiler, 1962
Amblyomma gemma	Kenya	Walker, 1974
Amblyomma gemma	Tanzania	Yeoman & Walker, 1967
Amblyomma gemma	Ethiopia	Theiler, 1962
Amblyomma gemma	New York Zoo	Becklund, 1968
Amblyomma hebraeum	Ethiopia	Theiler, 1962
Amblyomma lepidum	Ethiopia	Theiler, 1962
Amblyomma lepidum	Sudan	Hoogstraal, 1956
Amblyomma sparsum	Ethiopia	Theiler, 1962
Amblyomma sparsum	Kenya	Walker, 1974
Amblyomma variegatum	Ethiopia	Theiler, 1962
Amblyomma variegatum	Kenya	Walker, 1974
Amblyomma variegatum	Sudan	Hoogstraal, 1956
Boophilus decoloratus	Ethiopia	Theiler, 1962
Boophilus decoloratus	Kenya	Walker, 1974
Boophilus decoloratus	Transvaal	Innis, 1958
Boophilus decoloratus	New York Zoo	Becklund, 1968
Hyalomma albiparmatum	Tanzania	Yeoman & Walker, 1967
Hyalomma albiparmatum	Kenya	Walker, 1974
Hyalomma albiparmatum	Transvaal	Innis, 1958
Hyalomma albiparmatum	New York Zoo	Becklund, 1968
Hyalomma marginatum rufipes	Ethiopia	Theiler, 1962
Hyalomma marginatum rufipes	Tanzania	Yeoman & Walker, 1967
Hyalomma marginatum rufipes	Kenya	Walker, 1974
Hyalomma marginatum rufipes	Sudan	Hoogstraal, 1956
Hyalomma marginatum rufipes	Namibia	Horak *et al.*, 1992
Hyalomma truncatum	Ethiopia	Theiler, 1962
Hyalomma truncatum	Kenya	Walker, 1974
Hyalomma truncatum	Sudan	Hoogstraal, 1956
Hyalomma truncatum	Zambia	Colbo, 1973
Hyalomma truncatum	Namibia	Horak *et al.*, 1992
Margaropus reidi	Ethiopia	Theiler, 1962
Margaropus reidi	Sudan	Hoogstraal, 1956

Margaropus wileyi	Kenya	Walker & Laurence, 1973	
Rhipicephalus appendiculatus	Ethiopia	Theiler, 1962	
Rhipicephalus appendiculatus	Kenya	Walker, 1974	
Rhipicephalus camelopardalis	Ethiopia	Theiler, 1962	
Rhipicephalus camelopardalis	Tanzania	Yeoman & Walker, 1967	
Rhipicephalus compositus	Tanzania	Yeoman & Walker, 1967	
Rhipicephalus evertsi	Ethiopia	Theiler, 1962	
Rhipicephalus evertsi evertsi	Kenya	Walker, 1974	
Rhipicephalus evertsi mimeticus	Namibia	Horak *et al.*, 1992	
Rhipicephalus hurti	Kenya	Walker, 1974	
Rhipicephalus longiceps	Namibia	Horak *et al.*, 1992	
Rhipicephalus longicoxatus	Kenya	Walker, 1974	
Rhipicephalus mühlensis	Ethiopia	Theiler, 1962	
Rhipicephalus mühlensis	Kenya	Walker, 1974	
Rhipicephalus neavei (= R. kochi)	Tanzania	Yeoman & Walker, 1967	
Rhipicephalus pravus	Ethiopia	Theiler, 1962	
Rhipicephalus pravus	Tanzania	Yeoman & Walker, 1967	
Rhipicephalus pravus	Kenya	Walker, 1974	
Rhipicephalus pulchellus	Kenya	Walker, 1974	
Rhipicephalus pulchellus	Ethiopia	Yeoman & Walker, 1967	
Rhipicephalus pulchellus	New York Zoo	Becklund, 1968	
Rhipicephalus senegalensis	Ethiopia	Theiler, 1962	
Rhipicephalus simus	Ethiopia	Hoogstraal, 1956	
Rhipicephalus simus	Kenya	Walker, 1974	
Rhipicephalus simus	Tanzania	Yeoman & Walker, 1967	
Rhipicephalus supertritus	Tanzania	Yeoman & Walker, 1967	
Rhipicephalus tricuspis	Ethiopia	Theiler, 1962	
Rhipicephalus tricuspis	Sudan	Hoogstraal, 1956	

Mites

Sarcoptes scabiei		Kenya	Alasaad *et al.*, 2012

Internal worms

Camelostrongylus mentulatus	?	Japanese zoo	Fukumoto *et al.*, 1996
Echinococcus sp.	Angolan	Namibia	Krecek *et al.*, 1990
Haemonchus mitchelli	Angolan	Namibia	Krecek *et al.*, 1990
Helminths	Cape	Spanish zoo	Garijo *et al.*, 2004
Monodontella giraffae*	Cape	Namibia	Bertelsen *et al.*, 2009
Monodontella giraffae	?	China, Tianjin Zoo	Ming *et al.*, 2010
Parabronema skrjabini	Angolan	Namibia	Krecek *et al.*, 1990
Skrjabinema spp.	Angolan	Namibia	Krecek *et al.*, 1990

Viruses

coronavirus, bovine-like	?	zoo	Hasoksuz *et al.*, 2007
herpesvirus	retic		Hoenerhoff *et al.*, 2006
herpesvirus	retic	zoo	Kasem *et al.*, 2008
pestivirus	?	?	Harasawa *et al.*, 2000
rotavirus	?	?	Mulerin *et al.*, 2008

Other pathogens

Babesia sp.**	Cape	South Africa	Oosthuizen *et al.*, 2009
Brucellosis	Angolan	Botswana	Alexander *et al.*, 2012
Cryptosporidium muris	Retic	Czech zoos	Kodàdkovà *et al.*, 2010

Cytauxzoon sp.	Angolan	Namibia	Krecek *et al.*, 1990
Mycoplasma sp.	Retic	Texas	Hammond *et al.*, 2003
Sarcocystis 3 spp.	Cape	South Africa	Bengis *et al.*, 1998
Theileria sp.**	Cape	South Africa	Oosthuizen *et al.*, 2009

Monodontella giraffae nematode worms were present in the left liver lobe area of seven wild-caught giraffe of race *giraffa*. Although these worms were thought to be lethal in captive giraffe, they seem instead to be a common parasite of giraffe and one that does not necessarily harm the liver (Bertelsen *et al.*, 2009).

**Blood specimens of five young adult giraffe from different parts of South Africa who died contained *Babesia* and *Theileria* piroplasm. It may have been these organisms that killed them (Oosthuizen *et al.*, 2009).

References

Alasaad, S., D. Ndeereh, L. Rossi, *et al.* 2012. The opportunistic *Sarcoptes scabiei:* A new episode from giraffe in the drought-suffering Kenya. *Veterinary Parasitology* **185**: 359–363.

Alexander, K. A., J. K. Blackburn, M. E. Vandewalle, *et al.* 2012. Buffalo, bush meat, and the zoonotic threat of brucellosis in Botswana. *PloS ONE* http://www.plosone.org/article/info:doi/10.1371/journal.pone.0032842

Allin, M. 1998. *Zarafa: A Giraffe's True Story from Deep in Africa to the Heart of Paris.* New York: Walker and Co.

Amube, N. J., M. Antonínová & K. Hillman-Smith. 2009. Giraffes of the Garamba National Park, Democratic Republic of Congo. *Giraffa* **3**,1: 8–10.

Anon., 2012a. Research study needed for giraffe disease in Ruaha National Park, Tanzania. *Giraffa* **6**,1: 24.

Anon., 2012b. Giraffe electrocuted by power cables at South African game reserve. *Giraffa* **6**,1: 22.

Anon., 2012c. Mercy killing for snared giraffe. *Giraffa* **6**,1: 14.

Anon., 2012d. New additions at Giraffe Manor/Center in Nairobi. *Giraffa* **6**,1: 20–21.

Anon., 2012e. Leg-lifting: A new fighting style documented among Masai and reticulated giraffe. *Giraffa* **6**,2: 17.

Anon., 2012f. Suspension of hunting by 2014, Republic of Botswana. *Giraffa* **6**,2: 20–21.

Anon., 2012g. Three-year-old giraffe suckling. *Giraffa* **6**,2: 21.

Anon., 2012h. Teaching some tricks. *Giraffa* **6**,2: 21.

Anthony, H. E. 1928–29. Horns and antlers, their evolution, occurrence and function in the Mammalia. *Bulletin of the New York Zoological Society* **31**: 179–216; 32: 3–24.

Anthony, R. & F. Coupin. 1925. Recherches anatomique sur l'okapi, *Okapia johnstoni* Scl. II. Les sinus et les cornets nasaux. *Revue Zoologique Africaine* **13**: 69–96.

Aprea, F., P. M. Taylor, A. Routh, *et al.* 2011. Spinal cord injury during recovery from anaesthesia in a giraffe. *Veterinary Record* **169**,2: 50B.

Aschaffenburg, R., M. E. Gregory, S. J. Rowland, *et al.* 1962. The composition of the milk of the giraffe (*Giraffa camelopardalis reticulata*). *Proceedings of the Zoological Society of London* **139**: 359–363.

Awange, J. L., O. Aseto & O. Ong'ang'a. 2004. A case study on the impact of giraffes in Ruma National Park in Kenya. *Journal of Wildlife Rehabilitation* **27**,2: 16–21.

Backhaus, D. 1959. Experimentelle Prüfung des Farbsehvermögens einer Masai-Giraffe (*Giraffa camelopardalis tippelskirchi* Matschie 1898). *Zeitschrift für Tierpsychologie* **16**: 468–477.

Backhaus, D. 1961. *Beobachtungen an Giraffen in Zoologischen Gärten und freier Wildbahn.* Bruxelles: Institut des Parcs National du Congo et du Ruanda-Urundi.

Badeer, H. S. 1988. Haemodynamics of the jugular vein in the giraffe. *Nature* **332**: 788–789.

Badeer, H. S. 1997. Is the flow in the giraffe's jugular vein a 'free' fall? *Comparative Biochemistry and Physiology* **118A**,3: 573–576.

Badlangana, N. L., J. W. Adams & P. R. Manger. 2009. The giraffe (*Giraffa camelopardalis*) cervical vertebral column: A heuristic example in understanding evolutionary processes? *Zoological Journal of the Linnean Society* **155**: 736–757.

Badlangana, N. L., J. W. Adams & P. R. Manger. 2011. A comparative assessment of the size of the frontal air sinus in the giraffe (*Giraffa camelopardalis*). *Anatomical Record* **294**: 931–940.

Badlangana, N. L., A. Bhagwandin, K. Fuxe, *et al.* 2007a. Observations on the giraffe central nervous system related to the corticospinal tract, motor cortex and spinal cord: What difference does a long neck make? *Neuroscience* **148**: 522–534.

Badlangana, N. L., A. Bhagwandin, K. Fuxe, *et al.* 2007b. Distribution and morphology of putative catecholaminergic and serotonergic neurons in the medulla oblongata of a sub-adult giraffe, *Giraffa camelopardalis*. *Journal of Chemical Neuroanatomy* **34**: 69–79.

Baer, D. J., O. T. Oftedal & G. C. Fahey. 1985. Feed selection and digestibility by captive giraffe. *Zoo Biology* **4**,1: 57–64.

Bagemihl, B. 1999. *Biological Exuberance: Animal Homosexuality and Natural Diversity*. New York, NY: St. Martin's Press.

Ball, R. L. 2010. Clinical issues associated with nutrition and feeding in managed giraffe. *Giraffa* **4**,1: 50.

Bashaw, M. J. 2011. Consistency of captive giraffe behavior under two different management regimes. *Zoo Biology* **30**: 371–378.

Bashaw, M. J., M. A. Bloomsmith, T. L. Maple, *et al.* 2007. The structure of social relationships among captive female giraffe (*Giraffa camelopardalis*). *Journal of Comparative Psychology* **121**,1: 46–53.

Bashaw, M. J., L. R. Tarou, T. S. Maki, *et al.* 2001. A survey assessment of variables related to stereotypy in captive giraffe and okapi. *Applied Animal Behaviour Science* **73**: 235–247.

Baxter, E. & A. B. Plowman. 2001. The effect of increasing dietary fibre on feeding, rumination and oral stereotypies in captive giraffes (*Giraffa camelopardalis*). *Animal Welfare* **10**: 281–290.

Beattie, A. J. 1985. *The Evolutionary Ecology of Ant–Plant Mutualisms*. Cambridge: Cambridge University Press.

Becklund, W. W. 1968. Ticks of veterinary significance found on imports in the United States. *Journal of Parasitology* **54**: 622–628.

Beddard, F. 1906. Description of the external characters of an unborn fetus of a giraffe (*G.c. wardi*). *Proceedings of the Zoological Society of London* **1906**: 463–468.

Ben Shaul, D. M. 1962. The composition of the milk of wild animals. *International Zoo Yearbook* **4**: 333–342.

Bengis, R. G., K. Odening, M. Stolte, *et al.* 1998. Three new *Sarcocystis* species, *Sarcocystis giraffae*, *S. klaseriensis*, and *S. camelopardalis* (Protozoa: Sarcocystidae) from the giraffe (*Giraffa camelopardalis*) in South Africa. *Journal of Parasitology* **84**,3: 562–565.

Bercovitch, F. B. 2013. Giraffe cow reaction to the death of her newborn calf. *African Journal of Ecology* **51**: 376–379.

Bercovitch, F. B., M. J. Bashaw & S. M. del Castillo. 2006. Sociosexual behavior, male mating tactics, and the reproductive cycle of giraffe *Giraffa camelopardalis*. *Hormones and Behavior* **50**: 314–321.

Bercovitch, F. B., M. J. Bashaw, C. G. Penny, *et al.* 2004. Maternal investment in captive giraffes. *Journal of Mammalogy* **85**,3: 428–431.

Bercovitch, F. B. & P. S. M. Berry. 2009a. Ecological determinants of herd size in the Thornicroft's giraffe of Zambia. *African Journal of Ecology* **48**: 962–971.

Bercovitch, F. B. & P. S. M. Berry. 2009b. Reproductive life history of Thornicroft's giraffe in Zambia. *African Journal of Ecology* **48**: 535–538.

Bercovitch, F. B. & P. S. M. Berry. 2013. Herd composition, kinship, and fission–fusion social dynamics among wild giraffe. *African Journal of Ecology* **51**: 206–216.

Bernstein, A. 2012. Training for chiropractic adjustments, cold laser therapy and stretching, to manage a giraffe with a cervical spine injury. *Giraffa* **6**,1: 30.

Berry, P. S. M. 1973. The Luangwa Valley giraffe. *Puku* (Zambia) **7**: 71–92.

Berry, P. S. M. 1978. Range movements of giraffe in the Luangwa Valley, Zambia. *East African Wildlife Journal* **16**: 77–83.

Berry, P. S. M. 2007. Notes: Unusual tactics used in browsing by the Luangwa giraffe, Zambia. *Giraffa* **1**,2: 15–16.

Berry, P. S. M. & F. B. Bercovitch. 2012. Darkening coat color reveals life history and life expectancy of male Thornicroft's giraffe. *Journal of Zoology, London* **287**,3: 157–232.

Bertelsen, M. F., K. Østergaard, J. Monrad, *et al.* 2009. *Monodontella* giraffae infection in wild-caught southern giraffes (*Giraffa camelopardalis giraffa*). *Journal of Wildlife Diseases* **45**,4: 1227–1230.

Biewener, A. A. 1983. Allometry of quadrupedal locomotion: The scaling of duty factor, bone curvature and limb orientation to body size. *Journal of Experimental Biology* **105**: 147–171.

Birkett, A. 2002. The impact of giraffe, rhino and elephant on the habitat of a black rhino sanctuary in Kenya. *African Journal of Ecology* **40**: 276–282.

Bligh, J. & A. M. Harthoorn. 1965. Continuous radiotelemetric records of the deep body temperature of some unrestrained African mammals under near-natural conditions. *Journal of Physiology* **176**: 145–162.

Boesch, C. 2009. *The Real Chimpanzee: Sex Strategies in the Forest.* Cambridge: Cambridge University Press.

Bolger, D. T., T. A. Morrison, B. Vance, *et al.* 2012a. Development and application of a computer-assisted system for photographic mark recapture analysis of giraffe populations. *Giraffa* **6**,1: 25.

Bolger, D. T., T. A. Morrison, B. Vance, *et al.* 2012b. A computer-assisted system for photographic mark–recapture analysis. *Methods in Ecology and Evolution* **3**,5: 813–822.

Bond, W. J. & D. Loffell. 2001. Introduction of giraffe changes acacia distribution in a South African savanna. *African Journal of Ecology* **39**: 286–294.

Borkowski, R., S. Citino, M. Bush, *et al.* 2009. Surgical castration of subadult giraffe (*Giraffa camelopardalis*). *Journal of Zoo and Wildlife Medicine* **40**,4: 786–790.

Borkowski, R., B. Irvine, J. Robertson, *et al.* 2010. Successful concurrent rearing of two giraffe calves by a single giraffe dam. *Giraffa* **4**,1: 42.

Bothman, J. D. P., N. van Rooyen & M. W. van Rooyen. 2004. Using diet and plant resources to set wildlife stocking densities in African savannas. *Wildlife Society Bulletin* **32**,3: 1–12.

Bouché, P., P.-C. Renaud, P. Lejeune, *et al.* 2009. Has the final countdown to wildlife extinction in Northern Central African Republic begun? *African Journal of Ecology* **48**: 994–1003.

Boué, C. 1970. Morphologie fonctionelle des dents labiales chez les ruminants. *Mammalia* **34**: 696–711.

Bourlière, F. 1961. Le sex-ratio de la girafe. *Mammalia* **25**: 467–471.

Bozkurt, M. F.., M. E. Alcıgır, N. Yomusak, *et al.* 2011. Hydatid cysts on lungs of two African giraffe. *Ankara Universitesi Veteriner Fakultesi* **58**,1: 65–67.

Brand, R. 2011a. Male coat colouration as a signal of status in giraffe. *Giraffa* **5**,1: 28.

Brand, R. 2011b. Male mating strategies among the giraffe of Etosha National Park. *Giraffa* **5**,1: 23.

Bredin, I. P., J. D. Skinner & G. Mitchell. 2008. Can osteophagia provide giraffes with phophorus and calcium? *Onderstepoort Journal of Veterinary Research* **75**: 1–9.

Brenneman, R. A. 2004. Subspecific assessment of giraffe in the North American captive population compared to extant giraffe populations across Africa using nuclear genetic marker analyses. Unpublished research report.

Brenneman, R. A., R. K. Bagine, D. M. Brown, *et al.* 2009a. Implications of closed ecosystem conservation management: The decline of Rothschild's giraffe (*Giraffa camelopardalis rothschildi*) in Lake Nakuru National Park, Kenya. *African Journal of Ecology* **47**,4: 711–719.

Brenneman, R. A., E. E. Louis & J. Fennessy. 2009b. Genetic structure of two populations of the Namibian giraffe, *Giraffa camelopardalis angolensis*. *African Journal of Ecology* **47**,4: 720–728.

Brink, C. 1954. Hendrik Hop's expedition to Great Namaqualand (1761). In E. Axelson, ed. *South African Explorers*. London: Oxford University Press.

Broman, I. 1938a. Über die ersten Entwicklungsstadien der Mähne der Giraffen und der Equidae. *Anatomischer Anzeiger* **85**: 241–249.

Broman, I. 1938b. Einige Erfahrungen aus einer Giraffenjagd. *Zoologische Garten* N.F. **10**: 84–94.

Brøndum, E., J. M. Hasenkam, N. H. Secher, *et al.* 2009. Jugular venous pooling during lowering of the head affects blood pressure of the anesthetized giraffe. *American Journal of Physiology, Integrated Comparative Physiology* **297**: R1058–R1065.

Brook, B. S. & T. J. Pedley. 2002. A model for time-dependent flow in (giraffe jugular) veins: Uniform tube properties. *Journal of Biomechanics* **35**: 95–107.

Brown, C. 2010. Historic distribution of giraffe in the Greater Fish River Canyon Complex, southern Namibia, and recommendations for reintroductions. *Giraffa* **4**,1: 31–34.

Brown, D. M. & R. A. Brenneman. 2006. Conserving evolutionary potential in the giraffe. *Giraffa* **1**,1: 5–7.

Brown, D. M., R. A. Brenneman, K.-P. Koepfli, *et al.* 2007. Extensive population genetic structure in the giraffe. *BMC Biology* **5**: 13 pp. http://creativecommons.org/licenses/by/2.0

Bru, A. 2012. How to train your giraffe? *Giraffa* **6**,1: 26–27.

Bryden, H. A. 1893. *Gun and Camera in Southern Africa*. London: Stanford.

Brynard, A. M. & U. de V. Pienaar. 1960. Annual report of the biologists, 1958/1959. *Koedoe* **3**: 1–205.

Bush, M. 2003. Giraffidae. In M. E. Fowler & R. E. Miller, eds. *Zoo and Wild Animal Medicine*, 5th ed. St. Louis, MO: Saunders.

Bush, M. & S. B. Citino. 2010. Review of current anesthesia procedures for captive and free-living giraffe. *Giraffa* **4**,1: 42.

Bush, M., R. S. Custer & J. C. Whitla. 1980. Hematology and serum chemistry profiles for giraffes (*Giraffa camelopardalis*): Variations with sex, age, and restraint. *Journal of Zoo Animal Medicine* **11**,4: 122–129.

Bush, M. & V. de Vos. 1987. Observations on field immobilization of free-ranging giraffe (*Giraffa camelopardalis*) using carfentanil and xylazine. *Journal of Zoo Animal Medicine* **18**,4: 135–140.

Bush, M., D. G. Grobler, J. P. Raath, *et al.* 2001. Use of medetomidine and ketamine for immobilization of free-ranging giraffes. *Journal of the American Veterinary Medical Association* **218**,2: 245–249.

Bux, F., A. Bhagwandin, K. Fuxe & P. R. Manger. 2010. Organization of cholinergic, putative catecholaminergic and serotonergic nuclei in the diencephalon, midbrain and pons of sub-adult male giraffes. *Journal of Chemical Neuroanatomy* **39**: 189–203.

Caister, L. E., W. M. Shields & A. Gosser. 2003. Female tannin avoidance: A possible explanation for habitat and dietary segregation of giraffes (*Giraffa camelopardalis peralta*) in Niger. *African Journal of Ecology* **41**: 201–210.

Cameron, E. Z. 2004. Facultative adjustment of mammalian sex ratios in support of the Trivers–Willard hypothesis: Evidence for a mechanism. *Proceedings of the Royal Society of London B* **271**: 1723–1728.

Cameron, E. Z. & J. T. du Toit. 2005. Social influences on vigilance behaviour in giraffes, *Giraffa camelopardalis*. *Animal Behaviour* **69**: 1337–1344.

Cameron, E. Z. & J. T. du Toit. 2007. Winning by a neck: Tall giraffes avoid competing with shorter browsers. *American Naturalist* **169**, 130–135.

Cano, I. & W. Pérez. 2009. Quantitative anatomy of the trachea of the giraffe (*Giraffa camelopardalis rothschildi*). *International Journal of Morphology* **27**,3: 905–908.

Cárcoba, M. N. H. 2012. Hand rearing of a giraffe born with flexor tendon laxity. *Giraffa* **6**,1: 34.

Carter, K. D. 2009. Social organization of giraffe populations. *Giraffa* **3**,1: 41–42.

Carter, K. D. 2011. Interesting giraffe behaviour in Etosha National Park. *Giraffa* **5**,1: 14.

Carter, K. D., R. Brand, *et al.* 2012a. Social networks, long-term associations and age-related sociability of wild giraffes. Unpublished paper.

Carter, K. D., C. H. Frère, *et al.* 2012b. Constraints on sociability: Seasonal changes in relationships among female giraffes *Giraffa camelopardalis*. Unpublished paper.

Carter, K. D., J. M. Seddon, J. K. Carter, *et al.* 2012c. Development of 11 microsatellite markers for *Giraffa camelopardalis* through 454 pyrosequencing, with primer options for an additional 458 microsatellites. *Giraffa* **6**,1: 36. (From *Conservation Genetics Resources*.)

Carter, K. D., J. M. Seddon, C. H. Frère, *et al.* 2013. Fission–fusion dynamics in wild giraffes may be driven by kinship, spatial overlap and individual social preferences. *Animal Behaviour* **85**,2: 385–394.

Casares, M., A. Bernhard, C. Gerique, *et al.* 2012. Hand-rearing Rothschild or Baringo giraffe *Giraffa camelopardalis* rothschildi calves at Bioparc Valencia, Spain, and Leipzig Zoo, Germany. *International Zoo Yearbook* **46**: 221–231.

Catalano, R., T. Bruckner, J. Gould, *et al.* 2005. Sex ratios in California following the terrorist attacks of September 11, 2001. *Human Reproduction* **20**,5: 1221–1227.

Cave, A. J. E. 1950. On the liver and gall-bladder of the giraffe. *Proceedings of the Zoological Society of London* **120**: 381–393.

Cawley, R. 1975. Hoof trimming in giraffes *Giraffa camelopardalis*. *International Zoo Yearbook* **15**: 227.

Chamaillé-Jammes, S., M. Valeix, M. Bourgarel, *et al.* 2009. Seasonal density estimates of common large herbivores in Hwange National Park, Zimbabwe. *African Journal of Ecology* **47**: 804–808.

Chamberlain, G. 2012. The procedure of male giraffe anaesthesia and neutering at Port Lympne Wild Animal Park, United Kingdom. *Giraffa* **6**,1: 26.

Chappell, R. & K. Seeburger. 2012. Training and husbandry for a giraffe with a long term disability resulting from an angular limb deviation. *Giraffa* **6**,1: 25.

Chase, M. J. 2011. Abundance and distribution of giraffe in northern Botswana. *Giraffa* **5**,1: 23–24.

Chege, S. M. 2012. Giraffe translocation from Aberdare Country Club to Sera Wildlife Conservancy. *Giraffa* **6**,1: 9–13.

Chicago Zoological Society. 2012. CZS Giraffe Enrichment. Report.

Christman, J. 2008. Giraffe body scoring. *Giraffa* **2**,1: 25.

Church, S. K. 2004. The giraffe of Bengal: A medieval encounter in Ming China. *Medieval History Journal* **7**,1: 1–37.

Churcher, C. S. 1978. Giraffidae. Chapter 25 in V. J. Maglio & H. B. S. Cooke, eds. *Evolution of African Mammals*, pp. 509–535. Cambridge, MA: Harvard University Press.

Ciofolo, I. 1995. West Africa's last giraffes: The conflict between development and conservation. *Journal of Tropical Ecology* **11**: 577–588.

Ciofolo, I. & Y. Le Pendu. 2002. The feeding behaviour of giraffe in Niger. *Mammalia* **66**,2: 183–194.

Citino, S. B. 2008. A simple field ventilator for large ungulates. *Giraffa* **2**,1: 23.

Citino, S. B., M. Bush, W. Lance, *et al.* 2006. Use of Thiafentanil (A3080), Meditomidine, and Ketamine for anesthesia of captive and free-ranging giraffe (*Giraffa camelopardalis*). *Proceedings of the American Association of Zoo Veterinarians* **2006**: 211–213.

Citino, S. B., M. Bush & L. G. Phillips. 1984. Dystocia and fatal hyperthermic episode in a giraffe. *Journal of the American Veterinary Medical Association* **185**, 11: 1440–1442.

Clauss, M., T. A. Franz-Odendaal, J. Brasch, *et al.* 2007. Tooth wear in captive giraffe (*Giraffa camelopardalis*): Mesowear analysis classifies free-ranging specimens as browsers but captive ones as grazers. *Journal of Zoo and Wildlife Medicine* **38**,3: 433–445.

Clauss, M., E. Kienzle & J. M. Hatt. 2003. Feeding practice in wild captive ruminants: Peculiarities in the nutrition of browsers/concentrate selectors and intermediate feeders. In A. Fidgett, M. Clauss, U. Gansloßer, *et al.*, eds. *Zoo Animal Nutrition II*, pp. 27–52. Fürth: Filander.

Clauss, M., M. Lechner-Doll, E. J. Flasch, *et al.* 2001. Comparative use of four different marker systems for the estimation of digestibility and low food intake in a group of captive giraffes (*Giraffa camelopardalis*). *Zoo Biology* **20**: 315–329.

Clauss, M., P. Rose, J. Hummel, *et al.* 2006. Serous fat atrophy and other nutrition-related health problems in captive giraffe (*Giraffa camelopardalis*) – An evaluation of 83 necropsy reports. European Association of Zoo and Wildlife Veterinarians, 6th scientific meeting, 24–28 May 2006, Budapest, Hungary.

Clauss, M., W. K. Suedmeyer & E. J. Flach. 1999. Susceptibility to cold in captive giraffe. *Proceedings of the American Association of Zoo Veterinarians* **1999**, 183–186.

Clerck, A, de. 1965. News from our parks. *African Wild Life* **19**: 34.

Cobbold, T. S. 1860. Contributions to the anatomy of the giraffe. *Proceedings of the Zoological Society of London* **1860**: 99–105.

Coe, M. J. 1967. 'Necking' behaviour in the giraffe. *Journal of Zoology, London* **151**: 313–321.

Colbert, E. H. 1935. The classification and the phylogeny of the Giraffidae. *American Museum Novitates* **800**: 1–15.

Colbert, E. H. 1938. The relationships of the okapi. *Journal of Mammalogy* **19**: 47–64.

Colbo, M. H. 1973. Ticks of Zambian wild animals: A preliminary checklist. *Puku (Zambia)* **7**: 97–105.

Cooper, S. 2010, Feb 5. Fort Langley zoo giraffe dies during hoof-trimming procedure, charges likely. *The Province* (Vancouver).

Cornelius, A. J., L. H. Watson & A. G. Schmidt. 2012. The diet of giraffe (*Giraffa camelopardalis*) on a wildlife ranch in the mosaic thicket of the southern Cape, South Africa. Unpublished report.

Cotter, C., J. Miller & T. Burger. 2012. Yesterday, today and tomorrow: The evolution of a giraffe care and training program. *Giraffa* **6**,1: 32.

Cracknell, J., L. Dalrymple & R. Pizzi. 2010. Postmortem evaluation of left flank laparoscopic access in an adult female giraffe (*Giraffa camelopardalis*). *Veterinary Medicine International Annual*.

Cramer, M. D. & A. D. Mazel. 2007. The past distribution of giraffe in KwaZulu-Natal. *South African Journal of Wildlife Research* **37**,2: 197–201.

Crandall, L. S. 1964. *The Management of Wild Mammals in Captivity*. Chicago, IL: University of Chicago Press.

Cranfield, M., M. A. Eckhaus, B. A. Valentine, et al. 1985. Listeriosis in Angolan giraffes. *Journal of the American Veterinary Medical Association* **187**,11: 1238–1240.

Crawford, T. 2012. Giraffes in zoos may lack enough fat for cold weather. *Giraffa* **6**,2: 19–20.

Crile, G. 1941. *Intelligence, Power and Personality*. New York, NY: McGraw-Hill.

Cully, W. 1958. *Giraffa camelopardalis*. *Parks and Recreation* **41**: 197–198.

Dagg, A. I. 1959. Food preferences of the giraffe. *Proceedings of the Zoological Society of London* **135**,4: 640–642.

Dagg, A. I. 1960. Gaits of the giraffe and okapi. *Journal of Mammalogy* **41**: 282.

Dagg, A. I. 1962a. The distribution of the giraffe in Africa. *Mammalia* **26**: 497–505.

Dagg, A. I. 1962b. The role of the neck in the movements of the giraffe. *Journal of Mammalogy* **43**: 88–97.

Dagg, A. I. 1962c. The subspeciation of the giraffe. *Journal of Mammalogy* **43**: 550–552.

Dagg, A. I. 1963. A French giraffe. *Frontiers* **27**: 115–117f.

Dagg, A. I. 1965. Sexual differences in giraffe skulls. *Mammalia* **29**: 610–612.

Dagg, A. I. 1968. External features of giraffe. *Mammalia* **32**: 657–669.

Dagg, A. I. 1970a. Preferred environmental temperatures of some captive mammals. *International Zoo Yearbook* **10**: 127–130.

Dagg, A. I. 1970b. Tactile encounters in a herd of captive giraffe. *Journal of Mammalogy* **51**: 279–287.

Dagg, A. I. 1971. *Giraffa camelopardalis*. *Mammalian Species No. 5*, pp. 1–8. American Society of Mammalogists.

Dagg, A. I. 1974. The locomotion of the camel (*Camelus dromedarius*). *Journal of Zoology* **174**: 67–78.

Dagg, A. I. 1978. *Camel Quest: Summer Research on the Saharan Camel*. Waterloo, ON: Otter Press.

Dagg, A. I. 1979. The walk of large quadrupedal mammals. *Canadian Journal of Zoology* **57**: 1157–1163.

Dagg, A. I. 1984. Homosexual behaviour and female–male mounting in mammals – A first survey. *Mammal Review* **14**: 155–185.

Dagg, A. I. 2006. *Pursuing Giraffe: A 1950s Adventure*. Waterloo, ON: Wilfrid Laurier University Press.

Dagg, A. I. 2009. *The Social Behavior of Older Animals*. Baltimore, MD: Johns Hopkins University Press.

Dagg, A. I. 2011. *Animal Friendships*. Cambridge: Cambridge University Press.

Dagg, A. I. & A. De Vos. 1968a. The walking gaits of some species of Pecora. *Journal of Zoology, London* **155**: 103–110.

Dagg, A. I. & A. De Vos. 1968b. Fast gaits of pecoran species. *Journal of Zoology, London* **155**: 499–506.

Dagg A. I. & J. B. Foster. 1976. *The Giraffe: Its Biology, Behavior, and Ecology*. New York, NY: Van Nostrand Reinholt.

Dagg, A. I. & J. B. Foster. 1982. *The Giraffe: Its Biology, Behavior, and Ecology*, enlarged edition. Malabar, FL: Robert E. Krieger.

Damen, M. 2008. Giraffes in Europe. *Giraffa* **2**,1: 8–14.

Damen, M. 2010. Letter from the outgoing EAZA giraffe EEP coordinator. *Giraffa* **4**,1: 38.

Damian, A., A. Gudea, I. Irimescu, *et al.* 2012. Comparative studies of the thoracic appendicular skeleton in the giraffe and in the cow. *Lucrari Stiintifice Medicina Veterinara* **45**, 1.

Davis, M. R., J. N. Langan, N. D. Mylniczenko, *et al.* 2009. Colonic obstructions in three captive reticulated giraffe (*Giraffa camelopardalis reticulata*). *Journal of Zoo and Wildlife Medicine* **40**,1: 181–188.

Deacon, F. 2012. Update from the field. *Giraffa* **6**,2: 11.

Deacon, F., T. C. van Wyk & G. N. Smit. 2012. Movement patterns and impact of giraffe (*Giraffa camelopardalis*) on the woody plants of a fenced area in the central Free State. *Giraffa* **6**,2: 22.

Dean, K. 2012. Caught on tape: Camera trapping giraffe in Etosha National Park, Namibia. *Giraffa* **6**,1: 15.

Deka, B. C., K. C. Nath & B. N. Borgohain. 1980. Clinical note on the morphology of the placenta in a giraffe (*Giraffa camelopardalis*). *Journal of Zoo Animal Medicine* **11**,4: 117–118.

del Castillo, S. M., M. J. Bashaw, M. L. Patton, *et al.* 2005. Fecal steroid analysis of female giraffe (*Giraffa camelopardalis*) reproductive condition and the impact of endocrine status on daily time budgets. *General and Comparative Endocrinology* **141**: 271–281.

Dell, L., N. Patzka, A. Bhagwandin, *et al.* 2012. Organization and number of orexinergic neurons in the hypothalamus of two species of Cetartiodactyla: A comparison of giraffe (*Giraffa camelopardalis*) and harbour porpoise (*Phocoena phocoena*). *Journal of Chemical Neuroanatomy* **44**,2: 98–109.

deMaar, T. W. 2009. Editorial. *Giraffa* **3**,2: 17.

Derscheid, J. M. & H. Neuville. 1925. Recherches anatomiques sur l'okapi, *Okapia johnstoni* Scl III. La rate. *Revue Zoologique Africaine* **13**: 97–101.

Desbiez, A. & K. Leus. 2009a. Population viability analysis of the population of West-African giraffes (*Giraffa camelopardalis peralta*) in Niger. *Giraffa* **3**,1: 25–28.

Desbiez, A. & K. Leus. 2009b. Population and habitat viability assessment (PHVA) workshop for the West-African giraffe subspecies (*Giraffa camelopardalis peralta*). *Giraffa* **3**,1: 2–5.

Dikeman, C., D. Pogge & 4 others. 2009. Influence of diet on serum chemistry values in captive giraffe over four years. *Giraffa* **3**,2: 18.

Dillon, J. & J. Bredahl. 2010. Keeper assisted rearing of a giraffe calf. *Giraffa* **4**,1: 51.

Dobroruka, L. J. 1976. Hair whorls in the reticulated giraffe. *Giraffa camelopardalis reticulata*. *Vestnik Ceskoslovenske Spolecnosti Zoologicke* **40**: 255–258.

Doherty, J. 2008. Why do giraffes occur in aggregated dispersion patterns? *Giraffa* **2**,1: 20.

Doherty, J. 2011. Reticulated Giraffe Project. Report. Samburu National Reserve, Kenya.

Doherty, J., R. Elwood & M. Scantlebury. 2012. Reticulated giraffe: The behavioural and population ecology of a disappearing megaherbivore. *Giraffa* **6**,1: 28.

Doherty, J., L. Hiby, R. Elwood, *et al.* 2011. ExtractCompare: Automated 3D pattern recognition. *Giraffa* **5**,1; 24.

Dumonceaux, G. A., J. E. Bauman & G. R. Camilo. 2006. Evaluation of progesterone levels in feces of captive reticulated giraffe (*Giraffa camelopardalis reticulata*). *Journal of Zoo and Wildlife Medicine* **37**,3: 255–261.

Dunning, E. 2010. The Cincinnati Zoo and Botanical Garden's Masai giraffe hand feeding program: From concept to success. *Giraffa* **4**,1: 49.

Du Toit, J. T. 1990a. Giraffe feeding on Acacia flowers: Predation or pollination? *African Journal of Ecology* **28**: 63–68.

Du Toit, J. T. 1990b. Feeding-height stratification among African browsing ruminants. *African Journal of Ecology* **28**: 55–61.

Eagleson, A. 2010. Implementing a safe and healthy public giraffe feeding program. *Giraffa* **4**,1: 51–2.

East, R. 1999. African Antelope Database 1998. IUCN/SSC Antelope Specialist Group. IUCN, Gland, Switzerland and Cambridge, UK.

EAZA Giraffe EEPs. 2006. *EAZA Husbandry and Management Guidelines for Giraffa camelopardalis*. Arnhem, the Netherlands: Burgers' Zoo.

Egglebusch, A. 2008. Giraffe husbandry and hoof health: What factors have significance? *Giraffa* **2**,1: 24.

Endo, H., D. Yamagiwas, M. Fukisawa, *et al.* 1997. Modified neck muscular system of the giraffe (*Giraffa camelopardalis*). *Annals of Anatomy* **179**: 481–485.

Enoch, N. 2012. So where's the deep end, then? *Giraffa* **6**,1: 22.

Epaphras, A. M., E. D. Karimuribo, D. G. Mpanduji, *et al.* 2012. Prevalence, disease description and epidemiological factors of a novel skin disease in giraffes (*Giraffa camelopardalis*) in Ruaha National Park, Tanzania. *Research Opinions in Animal and Veterinary Sciences* **2**,1: 60–65.

Epelu-Opio, J. 1975. The fine structure of the urinary bladder epithelium in the hartebeest, *Alcelaphus buselaphus cokii* and the Masai giraffe, *Giraffa camelopardalis tippelskirchi*. *Anatomische Anzeiger* **138**: 46–55.

Espinoza, E. O., B. W. Baker, T. D. Moores, *et al.* 2008. Forensic identification of elephant and giraffe hair artifacts using HATR FTIR spectroscopy and discriminant analysis. *Endangered Species Research* **9**: 239–246.

Fennessy, J. 2003. Palewinged starling gleaning on desert-dwelling giraffe, northwestern Namibia. *Bird Numbers* **12**,1: 20–21.

Fennessy, J. 2004. Ecology of desert-dwelling giraffe *Giraffa camelopardalis angolensis* in northwestern Namibia. PhD thesis. School of Biological Sciences, University of Sydney, Australia.

Fennessy, J. 2006. Ecology of the desert-dwelling giraffe *G.c. angolensis*. *Giraffa* **1**,1: 2–4.

Fennessy, J. 2008. An overview of giraffe *Giraffa camelopardalis* taxonomy, distribution and conservation status, with a Namibian comparative and focus on the Kunene region. *Journal Namibia Wissenschaftliche Gesellschaft*, **56**: 65–81.

Fennessy, J. 2009. Home range and seasonal movements of *Giraffa camelopardalis angolensis* in the northern Namib Desert. *African Journal of Ecology* **47**: 318–327.

Fennessy, J. 2011. Desert-life – Is it all about sun and sand? *Giraffa* **5**,1: 24–25.

Fennessy, J. 2012a. Giraffe Information Database (GiD). Giraffe Conservation Foundation. Windhoek, Namibia.

Fennessy, J. 2012b. Extant giraffe taxonomy: Statement from the IUCN SSC ASG International Giraffe Working Group. *Giraffa* **6**,1: 2.

Fennessy, J. 2012c. Obituary: Dr Jean-Patrick 'JP' Suraud. *Giraffa* **6**,2: 2–3.

Fennessy, J., F. Bock & three others. 2013. Mitochondrial DNA analyses show that Zambia's South Luangwa Valley giraffe (*Giraffa camelopardalis thornicrofti*) are genetically isolated. *African Journal of Ecology* online

Fennessy, J. & R. Brenneman. 2010. *Giraffa camelopardalis* ssp. rothschildi. In *IUCN Red List of Threatened Species*, Version 2012.2.

Fennessy, J., K. E. A. Leggett & S. Schneider. 2003. Distribution and status of the desert-dwelling giraffe (*Giraffa camelopardalis angolensis*) in northwestern Namibia. *African Zoology* **38**, 184–188.

Fernandez, L. T., M. J. Bashaw, R. L. Sartor, *et al.* 2008. Tongue twisters: Feeding enrichment to reduce oral stereotypy in giraffe. *Zoo Biology* **27**: 200–212.

Fischer, M. T., R. E. Miller & E. W. Houston. 1997. Serial tranquilization of a reticulated giraffe (*Giraffa camelopardalis reticulata*) using xylazine. *Journal of Zoo and Wildlife Medicine* **28**,2: 182–184.

FitzSimons, F. W. 1920. *The Natural History of South Africa. Mammals.* Vol. **3**. London: Longmans, Green.

Fleming, P. A., S. D. Hofmeyr, S. W. Nicolson, *et al.* 2006. Are giraffes pollinators or flower predators of *Acacia nigrescens* in Kruger National Park, South Africa? *Journal of Tropical Ecology* **22**: 247–253.

Flower, W. H. 1870. *An Introduction to the Osteology of the Mammalia.* London: Macmillan.

Forsythe, K. & L. Marker. 2011. Feeding ecology of giraffe (*Giraffa camelopardalis*) in a fenced game reserve in North-Central Namibia. *Giraffa* **5**,1: 25.

Foster, J. B. 1966. The giraffe of Nairobi National Park: Home range, sex ratios, the herd, and food. *East African Wildlife Journal* **4**: 139–148.

Foster, J. B. 1968. The biomass of game animals in Nairobi National Park, 1960–1966. *Journal of Zoology (London)* **155**: 413–425.

Foster, J. B. & A. I. Dagg. 1972. Notes on the biology of the giraffe. *East African Wildlife Journal* **10**: 1–16.

Foxworth, B. 1997, Jan. Live giraffe offspring by AI.*Veterinary Observer* **3**.

Frackowiak, H. & H. Jakubowski. 2008 Arterial vascularization in the giraffe brain. *Annales Zoologici Fennici* **45**,4: 353–359.

Frank, A. 1960. Die Glandula thyreoidea und die Glandulae thyreoideae accessoriae bei *Giraffa camelopardalis* L. *Acta Anatomica* **42**: 267.

Franz-Odendaal, T. A. 2004. Enamel hypoplasia provides insights into early systemic stress in wild and captive giraffes (*Giraffa camelopardalis*). *Journal of Zoology* **263**: 197–206.

Franz-Odendaal, T. A., A. Chinsamy & J. Lee-Thorp. 2004. High prevalence of enamel hypoplasia in an early Pliocene giraffid (*Sivatherium*) from South Africa. *Journal of Vertebrate Paleontology* **24**,1: 235–244.

Fukumoto, S., T. Uchida, M. Ohbayashi, *et al.* 1996. A new host record of *Camelostrongylus mentulatus* (Nematoda; Trichostrongyloidea) from abomasum of a giraffe at a zoo in Japan. *Journal of Veterinary Medical Science* **58**,12: 1223–1225.

Furstenburg, D. & W. van Hoven. 1994. Condensed tannin as anti-defoliate agent against browsing by giraffe (*Giraffa camelopardalis*) in the Kruger National Park. *Comparative Biochemistry and Physiology* **107A**,2: 425–431.

Fust, P. 2009. Influences of anthropogenic activities on the giraffe (*Giraffa camelopardalis*) population of Omo National Park. *Giraffa* **3**,1: 29–31,

Gage, L. J. 2012. Giraffe welfare in the United States. *Giraffa* **6**,1: 29.

Ganey, T., J. Ogden & J. Olsen. 1990. Development of the giraffe horn and its blood supply. *Anatomical Record* **227**: 497–507.

Gardner, H. M., B. L. Hull, J. A. Hubbell, *et al.* 1986. Volvulus of the ileum in a reticulated giraffe. *Journal of the American Veterinary Medical Association* **189**,9: 1180–1181.

Garijo, M. M., J. M. Ortiz & M. R. R. de Ibanez. 2004. Helminths in a giraffe (*Giraffa camelopardalis giraffa*) from a zoo in Spain. *Onderstepoort Journal of Veterinary Research* **71**,2: 153–156.

Garretson, P. D., E. E. Hammond, T. M. Craig, *et al.* 2009. Anthelmintic resistant *Haemonchus contortus* in a giraffe (*Giraffa camelopardalis*) in Florida. *Journal of Zoo and Wildlife Medicine* **40**,1: 131–139.

Gauthier-Pilters, H. & A. I. Dagg. 1981. *The Camel: Its Evolution, Ecology, Behavior, and Relationship to Man*. Chicago, IL: University of Chicago Press.

Gay, P. 2009. Twins – Bioparc Z00 Doué la fontaine. *Giraffa* **3**,1: 45–46.

Gaymer, J. 2011. Kenya's reticulated giraffe are fighting an uphill battle! *Giraffa* **5**,1: 16.

Geiser, D. R., P. J. Morris & H. S. Adair. 1992. Multiple anesthetic events in a reticulated giraffe (*Giraffa camelopardalis*). *Journal of Zoo and Wildlife Medicine* **23**,2: 189–196.

Geist, V. 1975. *Mountain Sheep and Man in the Northern Wilds*. Ithaca, NY: Cornell University Press.

Gilbert, D. E., N. M. Loskutoff, C. G. Dorn, *et al.* 1988. Hormonal manipulation and ultrasonographic monitoring of ovarian activity in the giraffe. *Theriogenology* **29**,1: 248.

Ginnett, T. F. & M. W. Demment. 1997. Sex differences in giraffe foraging behavior at two spatial scales. *Oecologia* **110**: 291–300.

Ginnett, T. & M. W. Demment. 1999. Sexual segregation by Masai giraffes at two spatial scales. *African Journal of Ecology* **37**: 93–106.

Glass, R. L., R. Jenness & L. W. Lohse. 1969. Comparative biochemical studies of milks. V. The triglyceride composition of milk fats. *Comparative Biochemistry and Physiology* **28**: 783–786.

Glover, T. D. 1973. Aspects of sperm production in some East African mammals. *Journal of Reproductive Fertility* **35**: 45–53.

Godina, A.Y. ca 1970. On the parallelism in the evolution of *Paleotragus* and *Giraffa* and its importance in some evolutionary studies (in Russian). *U.S.S.R. Academy of Sciences, Paleontological Institute Proceedings* **130**: 62–69.

Goetz, R. H. 1955. Preliminary observations on the circulation of the giraffe. *Transactions of the American College of Cardiology* **5**: 39–48.

Goetz, R. H. & E. N. Keen. 1957. Some aspects of the cardiovascular system in the giraffe. *Angiology* **8**: 542–564.

Gombe, S. & F. I. B. Kayanja. 1974. Ovarian progestins in Masai giraffe (*Giraffa camelopardalis*). *Journal of Reproductive Fertility* **40**,1: 45–50.

Goodman, A. H., G. J. Armelagos & J. C. Rose. 1980. Enamel hypoplasias as indicators of stress in three prehistoric populations from Illinois. *Human Biology* **52**,3: 515–528.

Granvik, H. 1925. Mammals from the eastern slopes of Mount Elgon, Kenya Colony. *Acta Universitatis Lundensis. N.S. Lunds Univ. Arsskrift* **21**: 1–36.

Greene, T. V., S. P. Manne & L. M. Reiter. 2006. Developing models for mother–infant behaviour in black rhinoceros and reticulated giraffe. *International Zoo Yearbook* **40**: 372–378.

Groo, M. 2012. South Sudan's elephants face extinction. http://www.france24.com/en/20121204-south-sudans-elephants-face-extinction-experts

Groves, C. & P. Grubb. 2011. *Ungulate Taxonomy*. Baltimore, MD: Johns Hopkins University Press.

Grunert, E., H. Bader, L. Dittrich, *et al.* 1988. Right fetlock flexion as a cause of dystocia in a giraffe. *Deutsche Tierarztliche Wochenschrift* **95**,8: 315–317.

Hagenbeck, C. H. 1960. Hufbehandlung einer erwachsenen Giraffe. *Der Zoologische Garten* **25**: 182–188.

Hall-Martin, A. J. 1974a. A note on the seasonal utilization of different vegetation types by giraffe. *South African Journal of Science* **70**: 122–123.

Hall-Martin, A. J. 1974b. Food selection by Transvaal lowveld giraffe as determined by analysis of stomach contents. *Journal of Southern African Wildlife Management Association* **4**,3: 191–202.

Hall-Martin, A. J. 1975. Studies on the biology and productivity of the giraffe *Giraffa camelopardalis*. DSc Thesis. University of Pretoria, South Africa.

Hall-Martin, A. J. 1976. Dentition and age determination of the giraffe *Giraffa camelopardalis*. *Journal of Zoology, London* **180**: 263–289.

Hall-Martin, A. J. 1977. Giraffe weight estimation using dissected leg weight and body measurements. *Journal of Wildlife Management* **41**: 740–745.

Hall-Martin, A. J. & W. D. Basson. 1975. Seasonal chemical composition of the diet of Transvaal lowveld giraffe. *Journal of the South African Wildlife Management Association* **5**: 19–21.

Hall-Martin, A. J. & I. W. Rowlands. 1980. Observations on the ovarian structure and development of the southern giraffe, *Giraffa camelopardalis giraffa*. *South African Journal of Zoology* **15**: 217–221.

Hall-Martin, A. J. & J. D. Skinner. 1978. Observations on puberty and pregnancy in female giraffe (*Giraffa camelopardalis*). *South African Journal of Wildlife Research* **8**,3: 91–94.

Hall-Martin, A. J., J. D. Skinner & A. Smith. 1977b. Observations on lactation and milk composition of the giraffe *Giraffa camelopardalis*. *South African Wildlife Research* **7**,2: 67–71.

Hall-Martin, A. J., J. D. Skinner & J. M. van Dyk. 1975. Reproduction in the giraffe in relation to some environmental factors. *East African Wildlife Journal* **13**: 237–248.

Hall-Martin, A. J., M. von la Chevallerie & J. D. Skinner. 1977. Carcass composition of the giraffe *Giraffa camelopardalis*. *South African Journal of Animal Science* **7**,1: 55–64.

Hammond, E. E., C. A. Miller, L. Sneed, *et al.* 2003. *Mycoplasma*-associated polyarthritis in a reticulated giraffe. *Journal of Wildlife Diseases* **39**,1: 233–237.

Hanström, B. 1953. The hypophysis in some South African Insectivora, Carnivora, Hyracoidea, Proboscidea, Artiodactyla and Primates. *Arkiv. Zool.* **4**,3: 187–294.

Harasawa, R., M. Giangaspero, G. Ibata, *et al.* 2000. Giraffe strain of pestivirus: Its taxonomic status based on the 5′-untranslated region. *Microbiology and Immunology* **44**,11: 915–921.

Hargens, A. R., R. W. Millard, K. Pettersson, *et al.* 1987. Gravitational haemodynamics and oedema prevention in the giraffe. *Nature* **329**: 59–60.

Harrison, D. F. N. 1980. Biomechanics of the giraffe larynx and trachea. *Acta Otolaryngologica* **89**: 258–264.

Harrison, D. F. N. 1981. Fibre size frequency in the recurrent laryngeal nerves of man and giraffe. *Acta Otolaryngologica* **91**: 383–389.

Hasoksuz, M., K. Alekseev, A. Vlasova, *et al.* 2007. Biologic, antigenic, and full-length genomic characterization of a bovine-like coronavirus isolated from a giraffe. *Journal of Virology* **81**,10: 4981–4990.

Hassanin, A., A. Ropiquet, A. L. Gourmand, *et al.* 2007. Mitochondrial DNA variability in *Giraffa camelopardalis*: Consequences for taxonomy, phylogeography and conservation of giraffes in West and central Africa. *Comptes Rendus Biologies* **330**,3: 265–274.

Hatt, J. M., M. Lechner-Doll & B. Mayes. 1998. The use of dosed and herbage *n*-alkanes as markers for the determination of digestive strategies of captive giraffes (*Giraffa camelopardalis*). *Zoo Biology* **17**,4: 295–309.

Hatt, J. M., D. Schaub, M. Wanner, *et al.* 2005. Energy and fibre intake in a group of captive giraffe (*Giraffa camelopardalis*) offered increasing amounts of browse. *Journal of Veterinary Medicine series A – Physiology Pathology Clinical Medicine* **52**,10:485–490.

Henderson, D. M. & D. Naish. 2010. Predicting the buoyancy, equilibrium and potential swimming ability of giraffes by computational analysis. *Journal of Theoretical Biology* **265**: 151–159.

Herzinger, T., E. Scharrer, M. Placzek, *et al.* 2005. Contact urticaria to giraffe hair. *International Archives of Allergy and Immunology* **138**,4: 324–327.

Hett, M. L. 1928. The comparative anatomy of the palatine tonsil. *Proceedings of the Zoological Society of London* **1928**: 843–915.

Hicks, J. W. & H. S. Badeer. 1989. Siphon mechanism in collapsible tubes: Application to circulation of the giraffe head. *American Journal of Physiology* **256**,2: R567–R571.

Hilsberg-Merz, S. 2008. Infrared thermography in zoo and wild animals. In M. E. Fowler & R. E. Miller, eds. *Zoo and Wild Animal Medicine*. Vol. **6**, pp. 20–32. St. Louis, MO: Saunders.

Hirst, S. M. 1969. Predation as a regulating factor of wild ungulate populations in a Transvaal lowveld nature reserve. *Zoological Africana* **4**: 199–230.

Hirst, S. M. 1975. Ungulate–habitat relationships in a South African woodland/savanna ecosystem. *Wildlife Monographs* **44**: 1–60.

Hoare, R. & S. Brown. 2010. Snaring, poaching and snare removal from giraffes in Serengeti, Tanzania. *Giraffa* **4**,1: 24–25.

Hoenerhoff, M. J., E. B. Janovitz, L. K. Richman, *et al.* 2006. Fatal herpesvirus encephalitis in a reticulated giraffe (*Giraffa camelopardalis reticulata*). *Veterinary Pathology* **43**,5: 769–772.

Hofmann, R. R. 1968. Comment. In M. A. Crawford, ed. *Comparative Nutrition of Wild Animals*, pp. 217–219. London: Academic Press.

Hoogstraal, H. 1956. *Ticks of the Sudan: African Ixodoidea*. Vol. **1**. Cairo, Egypt. US Naval Medical Research Unit 3, Research Report NM 005050.29.07. 1101 pp.

Horak, I. G., M. Anthonissen, R. C. Krecek, *et al.* 1992. Arthropod parasites of springbok, gemsbok, kudus, giraffes and Burchell's and Hartmann's zebras in the Etosha and Hardap Nature Reserves, Namibia. *Onderstepoort Journal of Veterinary Research* **59**: 253–257.

Hradecký, P. 1983a. Some uterine parameters in antelope and a giraffe. *Theriogenology* **20**,4: 491–498.

Hradecký, P. 1983b. Placental morphology in African antelopes and giraffes. *Theriogenology* **20**,6: 725–734.

Huang, L., A. Nesterenko, W. Nie, *et al.* 2008. Karyotype evolution of giraffes (*Giraffa camelopardalis*) revealed by cross-species chromosome painting with Chinese muntjac (*Muntiacus reevesi*) and human (*Homo sapiens*) paints. *Cytogenetic and Genome Research* **122**,2: 132–138.

Hugh-Jones, P., C. E. Barter , J. M. Hime, *et al.* 1978. Dead space and tidal volume of the giraffe compared with some other mammals. *Respiration Physiology* **35**: 53–58.

Hummel, J., W. Zimmermann, T. Langenhorst, *et al.* 2006. Giraffe husbandry and feeding practices in Europe: Results of an EEP Survey. European Association of Zoo and Wildlife Veterinarians, 6th scientific meeting, 24–28 May 2006, Budapest, Hungary.

Hussein, A. A. 2009. Community based giraffe conservation and poverty alleviation in Garissa, Kenya. *Giraffa* **3**,1: 11–13.

Iglesias, T. L., R. McElreath & G. L. Patricelli. 2012. Western scrub-jay funerals: Cacophonous aggregations in response to dead conspecifics. *Animal Behaviour* **84**,5, 1103–1111.

Innis, A. C. 1958. The behaviour of the giraffe, *Giraffa camelopardalis*, in the eastern Transvaal. *Proceedings of the Zoological Society of London* **131**: 245–278.

International Giraffe Working Group. 2012, Feb. Extant giraffe taxonomy: Statement from the IUCN SSC ASG.

Isobe, N., T. Nakao, M. Shimada, *et al.* 2007. Fecal progestagen and estrone during pregnancy in a giraffe: A case report. *Journal of Reproduction and Development* **53**,1: 159–164.

Jacobson, E. R., P. Poulos, K. Quesenberry, *et al.* 1986. Polyarthritis and polyosteomyelitis in a juvenile giraffe. *Journal of the American Veterinary Medical Association* **189**,9: 1182–1183.

James, S. B., K. Koss, J. Harper, *et al.* 2000. Diagnosis and treatment of a fractured third phalanx in a Masai giraffe (*Giraffa camelopardalis tippelskirchi*). *Journal of Zoo and Wildlife Medicine* **31**,3: 400–403.

Johnson, M. 1928. *Safari*. New York, NY: Putnam and Sons.

Juan-Salles, C., G. Martinez, M. M. Garner, *et al.* 2008. Fatal dystocia in a giraffe due to a pelvic chondrosarcoma. *Veterinary Record* **162**,11: 349–351.

Jubb, R. A. 1970. Animals killed by severe frost in Rhodesia. *African Wild Life* **24**: 241–243.

Kagaruki, L. K., R. Fyumagwa & 2 others. 2003, Dec 4–6. A study of giraffe ear problems in Mikumi National Park: Tick and fly survey. In *Proceedings of the 4th Annual Scientific Conference*, pp. 178–180. Arusha, Tanzania: Tanzania Wildlife Research Institute.

Kaitho, T. D., C. K. Limo, B. Rono, *et al.* 2011. Dystocia in a Rothschild giraffe at the African Fund for Endangered Wildlife, Nairobi, Kenya. *Veterinary World* **4**,12: 565–568.

Kandrac, A. 2011. Teaching young giraffe old tricks: Changing learned behaviors in a herd of captive giraffe. *Animal Keepers' Forum: Dedicated Issue on Ungulate Husbandry, Training, Enrichment and Conservation* **38**: 286–290.

Karesh, W. B. 2008. Biopsy darting. In M. E. Fowler & R. E. Miller, eds. *Zoo and Wild Animal Medicine: Current Therapy.* Vol. **6**, pp. 105–109. St. Louis, MO: Saunders.

Kasem, S., S. Yamada & M. Kiupel. 2008. Equine herpesvirus type 9 in giraffe with encephalitis. *Emerging Infectious Diseases* **14**,12: 1948–1949.

Kayanja, F. I. B. 1973. The ultrastructure of the mandibular and ventral buccal glands of some East African wild ungulates. *Anatomischer Anzeiger* **134**: 339–350.

Kayanja, F. I. B. & L. H. Blankenship. 1973. The ovary of the giraffe, *Giraffa camelopardalis*. *Journal of Reproductive Fertility* **34**: 305–313.

Kayanja, F. I. B. & P. Scholz. 1974. The ultrastructure of the parotid gland of some East African wild ungulates. *Anatomischer Anzeiger* **135**,4: 382–397.

Keen, E. N. & R. H. Goetz. 1957. Cardiovascular anatomy of a foetal giraffe. *Acta Anatomica* **31**: 562–571.

Kellas, L. M., E. W. van Lennep & E. C. Amoroso. 1958. Ovaries of some foetal and prepubertal giraffes (*Giraffa camelopardalis*, Linnaeus). *Nature* **181**,4607: 487–488.

Kempter, C., J. Maltzan, H. Gerhards, *et al.* 2009. Bilateral patellafixation in a sub-adult reticulated giraffe (*Giraffa camelopardalis reticulata*). *Giraffa* **3**,2: 18.

Kidd, W. 1900. The significance of the hair-slope in certain mammals. *Proceedings of the Zoological Society of London* **1900**: 676–686.

Kidd, W. 1903. Traces of animal habits. In *Hutchinson's Animal Life*, pp. 234–235. London: Hutchinson.

Kimani, J. K. 1981a. Structural evidence of insertion of collagen fibres to smooth muscle cells in the carotid arterial system of the giraffe (*Giraffa camelopardalis*). *Cell and Tissue Research* **214**: 219–224.

Kimani, J. K. 1981b. Sub-endothelial fibrillar laminae in the carotid arteries of the giraffe (*Giraffa camelopardalis*). *Cell and Tissue Research* **219**,2: 441–443.

Kimani, J. K. 1983a. The structural organization of the carotid arterial system of the giraffe (*Giraffa camelopardalis*). *African Journal of Ecology* **21**: 317–324.

Kimani, J. K. 1983b. The structural organization of the tunica intima in the carotid arteries of the giraffe (*Giraffa camelopardalis*). *African Journal of Ecology* **21**: 309–315.

Kimani, J. K. 1987. Structural organization of the vertebral artery in the giraffe (*Giraffa camelopardalis*). *Anatomical Record* **217**,3: 256–262.

Kimani, J. K., R. N. Mbuva & R. M. Kinyamu. 1991a. Sympathetic innervation of the hindlimb arterial system in the giraffe (*Giraffa camelopardalis*). *Anatomical Record* **229**,1: 103–108.

Kimani, J. K. & J. M. Mungai. 1983. Observations on the structure and innervation of the presumptive carotid-sinus area in the giraffe (*Giraffa camelopardalis*). *Acta Anatomica* **115**,2: 117–133.

Kimani, J. K. & I. O. Opole. 1991. The structural organization and adrenergic-innervation of the carotid arterial system of the giraffe (*Giraffa camelopardalis*). *Anatomical Record* **230**,3: 369–377.

Kimani, J. K., I. O. Opole & J. A. Ogengo. 1991b. Structure and sympathetic innervation of the intracranial arteries in the giraffe (*Giraffa camelopardalis*). *Journal of Morphology* **208**,2: 193–203.

Kinney, A. 2008. Copper oxide wire particles for controlling intestinal parasites in giraffe. *Giraffa* **2**,1: 23.

Kinney-Moscona, A., D. K. Fontenot, J. E. Oosterhuis, *et al.* 2009. Variations in gastrointestinal parasites in multiple hoofstock species in different zoological facilities. *Giraffa* **3**,2: 19.

Kisling, V. N. Jr. 2001. *Zoo and Aquarium History: Ancient Animal Collections to Zoological Gardens.* Boca Raton, FL: CRC Press.

Kladetzky, J. 1954. Über Morphologie und Lage der Löwen- und Giraffenhypophyse. *Anatomischer Anzeiger* **100**: 202–216.

Klasen, S. A. H. 1963. Giraffe are strictly browsers. *Black Lechwe* **3**,5: 23–25.

Kleynhans, C. J. & W. van Hoven. 1976. Rumen protozoa of the giraffe with a description of two new species. *East African Wildlife Journal* **14**: 203–214.

Kobara, J. & T. Kamiya. 1965. On the gall-bladder of the giraffe (in Japanese). *Acta Anatomica Nipponica* **40**: 161–165.

Kodádková, A., M. Kváč, O. Ditrich, *et al.* 2010. *Cryptosporidium muris* in a reticulated giraffe (*Giraffa camelopardalis reticulata*). *Journal of Parasitology* **96**,1: 211–212.

Kodric-Brown, A. & J. H. Brown. 1984. Truth in advertising: The kinds of traits favored by sexual selection. *American Naturalist* **124**,3: 309–323.

Kok, O. B. 1982. Cannibalism and post-partum return to oestrus of a female Cape giraffe. *African Journal of Ecology* **20**: 141–143.

Kok, O. B. & D. P. J. Opperman. 1980. Feeding behavior of giraffe *Giraffa camelopardalis* in the Willem Pretorius Game Reserve, Orange Free State. *South African Journal of Wildlife Research* **10**,2: 45–55.

Koulischer, L., J. Tijskens & J. Mortelmans. 1971. Mammalian cytogenetics: V. The chromosomes of a female giraffe. *Acta Zoologica Pathologica Antverpiensia* **52**: 93–95.

Koutsos, E. A., D. Armstrong, R. Ball, *et al.* 2011. Influence of diet transition on serum calcium and phosphorus and fatty acids in zoo giraffe (*Giraffa camelopardalis*). *Zoo Biology* **30**: 523–531.

Krecek, R. C., J. Boomker, B. L. Penzhorn, *et al.* 1990. Internal parasites of giraffes (*Giraffa camelopardalis angolensis*). *Journal of Wildlife Diseases* **26**,3: 395–397.

Kristal, M. B. & M. Noonan. 1979a. Perinatal maternal and neonatal behaviour in the captive reticulated giraffe. *South African Journal of Zoology* **14**: 103–107.

Kristal, M. B. & M. Noonan. 1979b. Note on sleep in captive giraffes (*Giraffa camelopardalis reticulata*). *South Africa Journal of Zoology* **14**: 108.

Krumbiegel, I. 1939. Die Giraffe. Unter besonderer Berücksichttigung der Rassen. *Monogr. Wildsäugetiere* **8**: 1–98.

Krumbiegel, I. 1965. Gabelungsspuren an Giraffenhörnern. *Säugetierkundliche Mitteilungen* **13**: 107–108.

Kruuk, H. & M. Turner. 1967. Comparative notes on predation by lion, leopard, cheetah and wild dog in the Serengeti area, East Africa. *Mammalia* **31**: 1–27.

Kuntsch, V. & J. A. J. Nel. 1990. Possible thermoregulatory behaviour in *Giraffa camelopardalis*. *Zeitschrift für Säugetierkunde* **15**: 306–311.

Lahkar, K., M. Bhattacharya & A. Chakraborty. 1992. Histomorphological study on the tongue of giraffe. *Indian Veterinary Journal* **69**, 523–526.

Lahkar, K., A. Chakraborty & M. Bhattacharya. 1989. Gross anatomical observations on the tongue of giraffe. *Indian Veterinary Journal* **66**,10: 954–957.

Lamberski, N. & A. Blue. 2010. Preparing to anesthetize a giraffe in a confined area. *Giraffa* **4**,1: 52.

Lamprey, H. F. 1963. Ecological separation of the large mammal species in the Tarangire Game Reserve, Tanganyika. *African Journal of Ecology* **1**,1: 63–92.

Lamprey, H. F. 1964. Estimation of the large mammal densities, biomass and energy exchange in the Tarangire Game Reserve and Masai Steppe of Tanganyika. *East African Wildlife Journal* **2**: 1–46.

Lang, E. M. 1955. Beobachtungen während zweier Giraffengeburten. *Säugetierkundliche Mitteilungen* **3**,1: 1–5.

Langman, V. A. 1973. The immobilization and capture of giraffe. *South African Journal of Science* **69**: 200–203.

Langman, V. A. 1977. Cow–calf relationships in giraffe (*Giraffa camelopardalis giraffa*). *Zeitschrift für Tierpsychologie* **43**: 264–286.

Langman, V. A. 1978. Giraffe pica behavior and pathology as indicators of nutritional stress. *Journal of Wildlife Management* **42**,1: 141–147.

Langman, V. A. 1982. Giraffe youngsters need a little bit of maternal love. *Smithsonian* **12**,10: 94–103.

Langman, V. A. 2012. Thermal adaptations in giraffe. 5 pp. report.

Langman, V. A., O. S. Bamford & G. M. O. Maloiy. 1982. Respiration and metabolism in the giraffe. *Respiration Physiology* **50**: 141–152.

Langman, V. A. & G. M. O. Maloiy. 1989. Passive obligatory heterothermy of the giraffe. *Journal of Physiology, London* **415**: 89.

Langman, V. A., G. M. O. Maloiy, K. Schmidt-Nielsen, *et al.* 1979. Nasal heat exchange in the giraffe and other large mammals. *Respiration Physiology* **37**: 325–333.

Lankester, E. R. 1902. On *Okapia*, a new genus of Giraffidae from central Africa. *Transactions of the Zoological Society of London* **16**: 279–314.

Lankester, E. R. 1907. Parallel hair-fringes and colour-striping on the face of foetal and adult giraffes. *Proceedings of the Zoological Society of London* **1907**: 115–125.

Lankester, E. R. 1910. *Monograph of the Okapi.* London: Trustees of British Museum.

Laufer, B. 1928. The giraffe in history and art. Chicago, IL: Field Museum of Natural History, Anthropology Leaflet 27, 100 pp.

Lawrence, W. E. & R. E. Rewell. 1948. The cerebral blood supply in the Giraffidae. *Proceedings of the Zoological Society of London* **118**: 202–212.

Lechowski, R., J. Pisarski, J. Gosławski, *et al.* 1991. Exocrine pancreatic insufficiency-like syndrome in giraffe. *Journal of Wildlife Diseases* **27**,4: 728–730.

Lécu, A. 2010. Giraffe hoof trimming techniques under general anesthesia. *Giraffa* **4**,1: 45–46.

Leggett, K., J. Fennessy & S. Schneider. 2004. A study of animal movement in the Hoanib River catchment, north-western Namibia. *African Zoology* **39**,1: 1–11.

Le Pendu, Y. & I. Ciofolo. 1999. Seasonal movements of giraffes in Niger. *Journal of Tropical Ecology* **15**: 341–353.

Le Pendu, Y., I. Ciofolo & A. Gosser. 2000. The social organization of giraffes in Niger. *African Journal of Ecology* **38**: 78–85.

Lee, A. H., S. C. Chin & C.-T. Lin. 2006. Corneal trauma in a giraffe (*Giraffa camelopardalis*). *Proceedings of the American Association of Zoo Veterinarians* **2006**: 112.

Lee, D. E. & M. Bond. 2011. A picture is worth a thousand words: Studying demography of Masai giraffe using photographic mark–recapture. *Giraffa* **5**,1: 4–5.

Leggett, K., J. Fennessy & S. Schneider. 2004. A study of animal movement in the Hoanib River catchment, northwestern Namibia. *African Zoology* **39**,1: 1–11.

Leroy, R., M.-N. de Visscher, O. Halidou, *et al.* 2009. The last African white giraffes live in farmers' fields. *Biodiversity Conservation* **18**: 2663–2677.

Leuthold, B. M. 1979. Social organization and behavior of giraffe (*Giraffa camelopardalis*) in Tsavo East National Park, Kenya. *African Journal of Ecology* **17**: 19–34.

Leuthold, B. M. & W. Leuthold. 1972. Food habits of giraffe in Tsavo National Park, Kenya. *East African Wildlife Journal* **10**: 129–141.

Leuthold, B. M. & W. Leuthold. 1978a. Ecology of the giraffe in Tsavo East National Park, Kenya. *East African Wildlife Journal* **16**: 1–20.

Leuthold, B. M. & W. Leuthold. 1978b. Daytime activity patterns of gerenuk and giraffe in Tsavo National Park, Kenya. *East African Wildlife Journal* **16**: 231–243.

Leuthold, W. 1971. Studies on food habits of lesser kudu in Tsavo National Park, Kenya. *East African Wildlife Journal* **9**: 35–45.

Leuthold, W. & B. M. Leuthold. 1975. Temporal patterns of reproduction in ungulates of Tsavo East National Park, Kenya. *East African Wildlife Journal* **13**: 159–169.

Liu, L. M., D. W. Leaman & R. M. Roberts. 1996. The interferon-tau genes of the giraffe, a nonbovid species. *Journal of Interferon and Cytokine Research* **16**,11: 949–951.

Lochte, T. 1952. Das mikroskopische Bild des Giraffenhaares. *Zoologische Garten* **19**: 204–206.

Long, J. L., M. A. Raghanti & P. M. Dennis. 2010. Examination into the use of serologic biomarkers Leptin, IGF-1 and insulin; Glucose in the assessment of giraffe health. *Giraffa* **4**,1: 48.

Long, L. J., J. St. Ledger, P. M. Dennis, *et al.* 2006. Descriptive statistics of captive giraffe (*Giraffa camelopardalis*) mortality in American zoo and aquarium-accredited facilities from 1988–2005. *Proceedings of the American Association of Zoo Veterinarians* **2006**: 55.

Loskutoff, N. M., L. Walker, J. E. Ott-Joslin, *et al.* 1986. Urinary steroid evaluations to monitor ovarian-function in exotic ungulates. 2. Comparison between the giraffe (*Giraffa camelopardalis*) and the okapi (*Okapia johnstoni*). *Zoo Biology* **5**,4: 331–338.

Lueders, I., T. B. Hildebrandt, J. Pootoolal, *et al.* 2009b. Ovarian ultrasonography correlated with fecal progestins and estradiol during the estrous cycle and early pregnancy in giraffes (*Giraffa camelopardalis*). *Biology of Reproduction* **81**: 989–995.

Lueders, I., C. Niemuller, J. Pootoolal, *et al.* 2009a. Sonomorphology of the reproductive tract in male and pregnant and non-pregnant female Rothschild's giraffes (*Giraffa camelopardalis rothschildi*). *Theriogenology* **72**: 22–31.

Lydekker, R. 1904. On the subspeciation of *Giraffa camelopardalis*. *Proceedings of the Zoological Society of London* **1904**, 1: 202–227.

Lydekker, R. 1911. Two undescribed giraffe. *Nature* **87**: 484.

Madden, D. & T. P. Young. 1992. Symbiotic ants as an alternative defense against giraffe herbivory in spinescent *Acacia drepanolobium*. *Oecologia* **91**: 235–238.

Madeira, P. C. 1909. *Hunting in British East Africa*. Philadelphia, PA: Lippincott.

Mainka, S. A. & R. M. Cooper. 1989. Hypothyroidism in a reticulated giraffe (*Giraffa camelopardalis reticulata*). *Journal of Zoo and Wildlife Medicine* **20**,2: 217–220.

Maluf, N. S. R. 2002. Kidney of giraffes. *Anatomical Record* **267**: 94–111.

Marais, A. J., S. Fennessy & J. Fennessy. 2012a. *Country Profile: A Rapid Assessment of the Giraffe Conservation Status in South Sudan*. Giraffe Conservation Foundation, Windhoek, Namibia.

Marais, A. J., S. Fennessy & J. Fennessy. 2012b. *Country Profile: A Rapid Assessment of the Giraffe Conservation Status in Ethiopia*. Giraffe Conservation Foundation, Windhoek, Namibia.

Marais, A. J., S. Fennessy & J. Fennessy. 2012c. *Country Profile: A Rapid Assessment of the Giraffe Conservation Status in Rwanda*. Giraffe Conservation Foundation, Windhoek, Namibia.

Marais, A. J., S. Fennessy & J. Fennessy. 2012d. Giraffe conservation status report – country profile: Democratic Republic of the Congo. *Giraffa* **6**,2: 13–16.

Marais, A. J., L. Watson & A. Schmidt. 2011. Management of extra-limital giraffe (*Giraffa camelopardalis giraffa*) in Mosaic Thicket in South Africa. *Giraffa* **5**,1: 29.

Marealle, W. N., F. Fossøy, T. Holmern, *et al.* 2010. Does illegal hunting skew Serengeti wildlife sex ratios? *Wildlife Biology* **16**: 419–429.

Margulis, J. 2008, Nov. Looking up. *Smithsonian Magazine* **39**: 36–44.

Martinez, G. 2010. Increasing the medical options: Giraffe training. *Giraffa* **4**,1: 43–44.

Mason, G. J. 1991. Stereotypies: A critical review. *Animal Behaviour* **41**: 1015–1037.

McCalden, T. A., D. Borsook, A. D. Mendelow, *et al.* 1977. Autoregulation and haemodynamics of giraffe carotid blood flow. *South African Journal of Science* **73**: 278–279.

McCartney, M. 2010. Starting from scratch: Building a new giraffe facility from the ground up. *Giraffa* **4**,1: 47.

Medić, S. 2012. Case report of per acute death of Rothschild giraffe in Belgrade Zoo. *Giraffa* **6**,1: 31.

Meinhardt, K. & E. Teravskis. 2010. Incorporating modern husbandry into a new giraffe barn at the Nashville Zoo. *Giraffa* **4**,1: 42–43.

Middleton, S., T. H. Herdt, D. Agnew, *et al.* 2009. Post-mortem nutritional evaluation of bone mineral concentrations in the horse, cow, and dog and its application to exotic species. *Giraffa* **3**,2: 18.

Midgley, J. J., P. McLean, M. Botha, *et al.* 2001. Why do some African thorn trees (*Acacia* spp.) have a flat-top: A grazer–plant mutualism hypothesis? *African Journal of Ecology* **39**: 226–228.

Milewski, A. V., T. P. Young & D. Madden. 1991. Thorns as induced defenses: Experimental evidence. *Oecologia* **86**: 70–75.

Millard, R. W., A. R. Hargens, K. Pettersson, *et al.* 1987. Blood pressures and intervalve distances in giraffe veins. *Federation Proceedings* **46**,3: 793.

Miller, M. 2012. Anything but normal giraffe births at the Toledo Zoo. *Giraffa* **6**,1: 25–26.

Miller, M., B. Coville, N. Abou-Madi, *et al.* 1999. Comparison of in vitro tests for evaluation of passive transfer of immunoglobulins in giraffe (*Giraffa camelopardalis*). *Journal of Zoo and Wildlife Medicine* **30**,1: 85–93.

Ming, Z., L. Zhang & L. Zhang. 2010. Redescription of *Monodontella giraffae* Yorke et Maplestone, 1926 (Nematoda, Ancylostomatidae) from a giraffe, *Giraffa camelopardalis*, from zoo in China, with a discussion on the taxonomic status of *Monodontella*. *Acta Parasitologica* **55**,1: 66–70.

McQualter, K. 2012. Botswana's first giraffe research. *Giraffa* **6**,2: 9.

Mitchell, G. 2009. The origins of the scientific study and classification of giraffe. *Transactions of the Royal Society of South Africa* **64**,1: 1–13.

Mitchell, G., J. P. Bobbitt & S. Devries. 2008. Cerebral perfusion pressure in giraffe: Modelling the effects of head-raising and -lowering. *Journal of Theoretical Biology* **252**: 98–108.

Mitchell, G., S. K. Maloney, D. Mitchell, *et al.* 2006. The origin of mean arterial and jugular venous blood pressures in giraffes. *Journal of Experimental Biology* **209**: 2515–2524.

Mitchell, G., D. Roberts, S. van Sittert, *et al.* 2013. Growth patterns and masses of the heads and necks of male and female giraffes. *Journal of Zoology* **290**,1: 49–57.

Mitchell, G. & J. D. Skinner. 1993. How giraffe adapt to their extraordinary shape. *Transactions of the Royal Society of South Africa* **48**,2: 207–218.

Mitchell, G. & J. D. Skinner. 2003. On the origin, evolution and phylogeny of giraffes *Giraffa camelopardalis*. *Transactions of the Royal Society of South Africa* **58**,1: 51–73.

Mitchell, G. & J. D. Skinner. 2004. Giraffe thermoregulation: A review. *Transactions of the Royal Society of South Africa* **59**,2: 109–118.

Mitchell, G. & J. D. Skinner. 2009. An allometric analysis of the giraffe cardiovascular system. *Comparative Biochemistry and Physiology* **154A**, 4: 523–529.

Mitchell, G. & J. D. Skinner. 2011. Lung volumes in giraffes, *Giraffa camelopardalis*. *Comparative Biochemistry and Physiology*, Part A **158**: 72–78.

Mitchell, G., O. L. van Schalkwyk & J. D. Skinner. 2005. The calcium and phosphorus content of giraffe (*Giraffa camelopardalis*) and buffalo (*Syncerus caffer*) skeletons. *Journal of Zoology, London* **267**: 55–61.

Mitchell, G., S. J. van Sittert & J. D. Skinner. 2009a. Sexual selection is not the origin of long necks in giraffes. *Journal of Zoology* **278**, 281–286.

Mitchell, G., S. J. van Sittert & J. D. Skinner. 2009b. The structure and function of giraffe jugular vein valves. *South African Journal of Wildlife Research* **39**,2: 175–180.

Mlengeya, T. D. K., M. M. A. Mtambo, *et al.* 2002, Dec 3–5. Severe infectious otitis in giraffes in Mikumi National Park and Selous Game Reserve, Tanzania. In *Proceedings of the 3rd Annual Scientific Conference, Arusha, Tanzania*, pp. 283–290. Arusha, Tanzania: Tanzania Wildlife Research Institute.

Mònico, M. & N. Ortega. 2012. Five giraffe successfully collared in Garamba National Park, DRC. *Giraffa* **6**,1: 4–5.

Mpanduji, D. G., A. Epaphras, *et al.* 2010. Immobilization and biochemical parameters in free ranging giraffe in Ruaha National Park, Tanzania. *Giraffa* **4**,1: 46.

Mulherin, E., J. Bryan, M. Beltman, *et al.* 2008. Molecular characterisation of a bovine-like rotavirus detected from a giraffe. *BMC Veterinary Research* **4**, article 46.

Muller, Z. 2010a. The curious incident of the giraffe in the night time. *Giraffa* **4**,1: 20–23.

Muller, Z. 2010b. Sticking our necks out: Developing a National Giraffe Conservation Strategy for Kenya. *Giraffa* **4**,1: 26–27.

Muller, Z. 2010c. The Rothschild's giraffe project, Kenya. *Giraffa* **4**,1: 37–38.

Muller, Z. 2011, Aug. Sticking their necks out. *Africa Geographic*. http://www.africageographic.com

Muller, Z. 2012a. Conservation of the endangered Rothschild's giraffe in Kenya. *Giraffa* **6**,1: 29.

Muller, Z. 2012b. Home at last. *Giraffa* **6**,2: 19.

Murai, A., T. Yanai, M. Kato, *et al.* 2007. Teratoma of the umbilical cord in a giraffe (*Giraffa camelopardalis*). *Veterinary Pathology* **44**,2: 204–206.

Murray, J. D. 1981. A pre-pattern formation mechanism for animal coat markings. *Journal of Theoretical Biology* **88**: 161–199.

Muse, E.A., see Epaphras, A. M. entry.

Nakakuki, S. 1983. The bronchial ramification and lobular division of the giraffe lung. *Anatomischer Anzeiger* **154**,4: 313–317.

Nelson, A. S. 2012. The way home: Moving a herd of 1.6 reticulated giraffe into a new facility. *Giraffa* **6**,2: 5–9.

Nesbit Evans, E.M. 1970. The reaction of a group of Rothschild's giraffe to a new environment. *East African Wildlife Journal* **8**: 53–62.

Neuville, M. H. 1935. L'urètre glandaire des girafes. *Bulletin Musée National de Histoire Natural*, Ser. **2**,7: 333–339.

Ngog, Nje, J. 1983. Structure et dynamique de la population de girafes du Parc National de Waza, Cameroun. *Revue Ecologique–la terre et la vie* **37**,1: 3–30.

Ngog Nje, J. 1984. Régime alimentaire de la girafe au Parc National de Waza, Cameroun. *Mammalia* **48**,2: 173–183.

Nilsson, O., J. S. Bööj, A. Dahlström, *et al.* 1988. Sympathetic innervation of the cardiovascular system in the giraffe. *Blood Vessels* **25**,6: 299–307.

Norton-Griffiths, M. 1979. The influence of grazing, browsing, and fire on the vegetation dynamics of the Serengeti. In A. R. E. Sinclair & M. Norton-Griffiths, eds. *Serengeti: Dynamics of an Ecosystem*, pp. 310–352. Chicago, IL: University of Chicago Press.

O'Kane, C. A. J., K. J. Duffy, B. R. Page, *et al.* 2011. Overlap and seasonal shifts in use of woody plant species amongst a guild of savanna browsers. *Journal of Tropical Ecology* **27**: 249–258.

Ombuor, J. 2011. Gentle giraffes face tall order in Garissa. *Giraffa* **5**,1: 11–12.

Omondi, P., R. Mayienda & M. Tchamba. 2009. Giraffe in Waza NP, Cameroon. *Giraffa* **3**,1: 42–43.

Oosthuizen, M. C., B. A. Allsopp, M. Troskie, *et al.* 2009. Identification of novel *Babesia* and *Theileria* species in South African giraffe (*Giraffa camelopardalis*, Linnaeus, 1758) and roan antelope (*Hippotragus equinus*, Desmarest 1804). *Veterinary Parasitology* **163**,1–2: 39–46.

Opio, H. 2012. Training progress of giraffes at UWEC achieved from Oakland and San Francisco Zoo USA. *Giraffa* **6**,1: 30–31.

Orban, D. A., R. J. Snider, J. M. Siegford, *et al.* 2012. *Effects of guest feeding programs on captive giraffe behavior.* Animal Behavior and Welfare Group, Michigan State University, poster.

Østergaard, K. H., M. F. Bertelsen, E.T. Brøndum, *et al.* 2011. Pressure profile and morphology of the arteries along the giraffe limb. *Journal of Comparative Physiology* B **181**: 691–698.

Owen, P. 2012, Jan 21. Is this really fun for all the family? *Mail Online.*

Owen, R. 1841. Notes on the anatomy of the Nubian giraffe. *Transactions of the Zoological Society of London* **2**: 217–248.

Owen-Smith, N. 2008. Changing vulnerability to predation related to season and sex in an African ungulate assemblage. *Oikos* **117**: 602–610.

Page, D. N. 2012. The height of a giraffe. *Giraffa* **6**,1: 36.

Parker, D. M. 2004. The feeding biology and potential impact of introduced giraffe (*Giraffa camelopardalis*) in the Eastern Cape Province, South Africa. Master's thesis. Rhodes University, South Africa.

Parker, D. M. & R. T. F. Bernard. 2005. The diet and ecological role of giraffe (*Giraffa camelopardalis*) introduced to the Eastern Cape, South Africa. *Journal of Zoology, London* **267**: 203–210.

Parker, D. M., R. T. F. Bernard & S. A. Colvin. 2003. The diet of a small group of extralimital giraffe. *African Journal of Ecology* **41**: 245–253.

Paton, J. F. R., C. J. Dickinson & G. Mitchell. 2009. Harvey Cushing and the regulation of blood pressure in giraffe, rat and man: Introducing 'Cushing's mechanism.' *Experimental Physiology* **94**,1: 11–17.

Patterson, C. B. 1977, Sep. Rescuing the Rothschild. *National Geographic*: 418–421.

Patton, M. L., M. J. Bashaw, S. M. del Castillo, *et al.* 2006. Long-term suppression of fertility in female giraffe using the GnRH agonist deslorelin as a long-acting implant. *Theriogenology* **66**: 431–438.

Pedley, T. J. 1987. How giraffes prevent oedema. *Nature* **329**: 13–14.

Pedley, T. J., B. S. Brook & R. S. Seymour. 1996. Blood pressure and flow rate in the giraffe jugular vein. *Philosophical Transactions of the Royal Society of London B* **351**: 855–866.

Pellew, R. A. 1983a. The giraffe and its food resource in the Serengeti: I. Composition, biomass and production of available browse. *African Journal of Ecology* **21**: 241–267.

Pellew, R. A. 1983b. The giraffe and its food resource in the Serengeti. II. Response of the giraffe population to changes in the food supply. *African Journal of Ecology* **21**: 269–283.

Pellew, R. A. 1983c. The impacts of elephant, giraffe and fire upon the *Acacia tortilis* woodlands of the Serengeti. *African Journal of Ecology* **21**: 41–74.

Pellew, R. A. 1984a. Food consumption and energy budgets of the giraffe. *Journal of Applied Ecology* **21**: 141–159.

Pellew, R. A. 1984b. The feeding ecology of a selective browser, the giraffe (*Giraffa camelopardalis tippelskirchi*). *Journal of Zoology* **202**: 57–81.

Pérez, W., M. Lima & M. Clauss. 2009a. Gross anatomy of the intestine in the giraffe (*Giraffa camelopardalis*). *Anatomia Histologia Embryologia (Journal of Veterinary Medicine)* **38**: 432–435.

Pérez, W., M. Lima, G. Pedrana, *et al.* 2009b. Heart anatomy of *Giraffa camelopardalis rothschildi*: A case report. *Veterinarni Medicina* **53**,3: 165–168.

Pérez, W., V. Michel, H. Jerbi, *et al.* 2012. Anatomy of the mouth of the giraffe (*Giraffa camelopardalis rothschildi*). *International Journal of Morphology* **30**,1: 322–329.

Perry, K. 2012. The usefulness of assigning individual targets to your giraffes and how to get started. *Giraffa* **6**,1: 31.

Peterson, C., H. Haefele & A. Eyres. 2010. Giraffe immobilizations in a semi-free-ranging environment at Fossil Rim Wildlife Center. *Giraffa* **4**,1: 52–53.

Peterson, H. 2012. Tail de-gloving incident and subsequent training for docking procedure. *Giraffa* **6**,1: 25.

Phelps, A. & L. Clifton-Bumpass. 2009. The advantages of proactive reinforcement training with captive giraffe. *Giraffa* **3**,1: 14–21.

Phelps, A. & L. Clifton-Bumpass. 2011. A perfect match: How training giraffe in the captive environment enables institutions to partner with and support conservation work. *Giraffa* **5**,1: 25–26.

Phelps, A., K. Emanuelson, A. Goodnight, *et al.* 2012. Treatment and recovery of traumatic septic arthritis of the fetlock joint of 1.0 reticulated giraffe. *Giraffa* **6**,1: 28–29.

Phelps, A. & S. Mellard. 2012. Can an eland and a giraffe be friends? Interesting behavioural observations at the Oakland Zoo. *Giraffa* **6**,1: 16–17.

Pienaar, U. de V. 1963. The large mammals of the Kruger National Park – their distribution and present-day status. *Koedoe* **6**: 1–37.

Pienaar, U. de V. 1969. Predator–prey relationships among the large mammals of the Kruger National Park. *Koedoe* **12**: 108–176.

Pocock, R. I. 1910. Cutaneous scent-glands of ruminants. *Proceedings of the Zoological Society of London* **1910**: 840–986.

Porterfield, J. 2012. *Giraffe enrichment.*Report. Oakland Zoo, Oakland, CA.

Potgieter, D. 2009. Dry season aerial total count, Zakouma National Park, Chad. *Giraffa* **3**,2: 25.

Potter, J. S. & M. Clauss. 2005. Mortality of captive giraffe (*Giraffa camelopardalis*) associated with serous fat atrophy: A review of five cases at Auckland Zoo. *Journal of Zoo and Wildlife Medicine* **36**,2: 301–307.

Powell, D. 2010. World's oldest giraffe. *Giraffa* **4**,1: 36.

Pratt, D. M. & V. H. Anderson. 1979. Giraffe cow–calf relationships and social development of the calf in the Serengeti. *Zeitschrift für Tierpsychologie* **51**: 233–251.

Pratt, D. M. & V. H. Anderson. 1982. Population, distribution, and behaviour of giraffe in the Arusha National Park, Tanzania. *Journal of Natural History* **16**: 481–489.

Pratt, D. M. & V. H. Anderson. 1985. Giraffe social behavior. *Journal of Natural History* **19**,4: 771–781.

Prothero, D. R. 2007. *Evolution: What the Fossils Say and Why it Matters*. New York, NY: Columbia University Press.

Putnam, B. 1947. *Animal X-rays*. New York, NY: Putnam.

Quesada, R., S. B. Citino, J. T. Easley, *et al*. 2011. Surgical resolution of an avulsion fracture of the peroneus tertius origin in a giraffe (*Giraffa camelopardalis reticulata*). *Journal of Zoo and Wildlife Medicine* **42**,2: 348–350.

Radcliffe, R. M., T. A. Turner, C. H. Radcliffe, *et al*. 1999. Arthroscopic surgery in a reticulated giraffe (*Giraffa camelopardalis reticulata*). *Journal of Zoo and Wildlife Medicine* **30**,3: 416–420.

Reichenbach, H. 1980. Carl Hagenbeck's Tierpark and modern zoological gardens. *Journal of the Society for the Bibliography of Natural History* **9**,4: 573–585.

Retterer, E. & H. Neuville. 1914. Du pénis et du gland d'une girafe. *Societie Biologique (Paris) Comptes Rendus* **77**: 499–501.

Rewell, R. E. 1987. Vascular system of the giraffe. *Nature* **329**: 589.

Roberts, A. 1951. *The Mammals of South Africa*. New York, NY: Hafner Publishing.

Roberts, A. 2009. The Brookfield Zoo giraffe program. *Giraffa* **3**,2: 6–7.

Rose, P. 2005. A comparison of feed intake and cause of death of captive giraffe in the United Kingdom. MSc thesis. Royal Veterinary College, London, UK.

Rose, P. 2008. Developments to the nutrition of captive giraffe (*Giraffa camelopardalis*). British Veterinary Zoological Society Abstract.

Rose, P. 2010. Do you have friends in high places? *Social attachment and partner preference in captive giraffe (*Giraffa camelopardalis*)*. 12th BIAZA Research Symposium, July, Chester Zoo, Chester, UK.

Rose, P. 2011. Knowsley Safari Park 40th Birthday Lecture: 'All about giraffe.' *Giraffa* **5**,1: 8–11.

Rose, P. 2012. Investigations in the social behaviour of captive giraffe: What do the animals tell us about their life in the zoo? *Giraffa* **6**,1: 33. (IAGCP Conference, Feb, San Francisco.)

Rose, P. & M. Clauss. 2006. A comparison of husbandry, feed intake and cause of death of captive giraffe (*Giraffa camelopardalis*) in the United Kingdom. *BIAZA Research Newsletter* **7**,3: 4–5.

Rose, P., J. Hummel & M. Clauss. 2006. Food and calculated energy intake of captive giraffe (*Giraffa camelopardalis*) in the UK. European Association of Zoo and Wildlife Veterinarians 6th scientific meeting, 24–28 May 2006, Budapest, Hungary

Rose, P. & S. Roffe. 2007. *The effect of lavender (*Lavandula angustifolia*) on the behaviour of captive giraffe (*Giraffa camelopardalis*)*. Report. Twycross Zoo, Warwickshire, UK.

Rose, P. & S. Roffe. 2010. Five years of giraffe-centred research at the East Midlands Zoological Society: Twycross Zoo. Where have we been and where do we go? *Giraffa* **4**,1: 48–49.

Rose, P., S. Roffe & M. Jermy. 2008. Enrichment methods for *Giraffa camelopardalis* and *Gazella dama mhorr* at the East Midland Zoological Society: Twycross Zoo. *RATAL* **35**,1: 19–24.

Rothschild, M. & H. Neuville. 1911. Recherches sur l'okapi et les girafes de l'est africain. *Annales Science Naturelle, Zoologie (Paris), 9th Ser.* **13**: 1–185.

Rouffignac, M. 1997, Jan. Nasal tube gives newborn giraffe a second chance. *Veterinary Observer* 1–2.

Roy, R. R., S. Graham & J. A. Peterson. 1988. Fiber type composition of the plantarflexors of giraffes (*Giraffa camelopardalis*) at different postnatal stages of development. *Comparative Biochemistry and Physiology* **91A**,2: 347–352.

Saito, M. & Gen'ichildani. 2011. The giraffes in the Katavi National Park, Tanzania – Social preference for group formation and for social behaviours. *Giraffa* **5**,1: 30.

Salifou, Z. 2012. Gestion concertée des ressources naturelles de la zone de concentration des girafes au Niger (PGCRN/ZCGN), l'ONG ATPF s'engage. *Giraffa* **6**,2: 12–13.

Samuels, D. 2012, June. Wild things: Animal nature, human racism, and the future of zoos. *Harper's*, 28–42.

Sarginson, J. & J. Nissen. 2012. Rothschild's giraffe translocation, Kigio Wildlife Conservancy. *Giraffa* **6**,2: 10.

Sasaki, M., H. Endo, H. Kogiku, *et al.* 2001. The structure of the masseter muscle in the giraffe (*Giraffa camelopardalis*). *Anatomia Histologia Embryologia – Journal of Veterinary Medicine Series C* **30**,5: 313–319.

Sathar, F., N. L. Badlangana & P. R Manger. 2010. Variations in the thickness and composition of the skin of the giraffe. *Anatomical Record* **293**,9: 1615–1627.

Sauer, J. J. C. 1983a. Food selection by giraffes in relation to changes in chemical composition of the leaves. *South African Journal of Science* **13**,1: 40–43.

Sauer, J. J. C. 1983b. A comparison between *Acacia* and *Combretum* leaves utilized by giraffe. *South African Journal of Animal Science* **13**,1: 43–44.

Sauer, J. J. C., J. D. Skinner & A. W. H. Neitz. 1982. Seasonal utilization of leaves by giraffes *Giraffa camelopardalis*, and the relationship of the seasonal utilization to the chemical composition of the leaves. *South African Journal of Zoology* **17**,4: 210–219.

Schillings, C. G. 1905. *With Flashlight and Rifle*. London: Hutchinson.

Schmidt, D. A., R. L. Ball, D. Grobler, *et al.* 2007. Serum concentrations of amino acids, fatty acids, lipoproteins, vitamins A and E, and minerals in apparently healthy, free-ranging southern giraffe (*Giraffa camelopardalis giraffa*). *Zoo Biology* **26**: 13–25.

Schmidt, D. A., R. B. Barbiers, M. R. Ellersieck, *et al.* 2011. Serum chemistry comparisons between captive and free-ranging giraffes (*Giraffa camelopardalis*). *Journal of Zoo and Wildlife Medicine* **42**,1: 33–39.

Schmidt, D. A., E. A. Koutsos, M. R. Ellersieck, *et al.* 2009. Serum concentration comparisons of amino acids, fatty acids, lipoproteins, vitamin A and E, and minerals between zoo and free-ranging giraffes (*Giraffa camelopardalis*). *Journal of Zoo and Wildlife Medicine* **40**: 29–38.

Schmidt-Nielsen, K. 1959. The physiology of the camel. *Scientific American* **201**,6: 140–151.

Schmidt-Nielsen, K., B. Schmidt-Nielsen, S. A. Jarnum, *et al.* 1957. Body temperature of the camel and its relation to water economy. *American Journal of Physiology* **188**: 103–112.

Schmucker, S. S., L. Kolter & G. Nogge. 2010. Analysis of the fine structure of oral activities in captivity unravels control of food intake in giraffe (*Giraffa camelopardalis*). *Giraffa* **4**,1: 49.

Schneider, K. M. 1951. Nachrichten aus Zoologischen Garten. *Zoologische Garten* **18**: 73.

Schreider, E. 1950. Geographical distribution of the body-weight/body-surface ratio. *Nature* **165**: 286.

Schultz, K., P. McNickle & L. Brown. 2010. The rogue thumb: Developing, evaluating and refining a volunteer-lead giraffe encounter. *Giraffa* **4**,1: 43.

Seeber, P. A., I. Ciofolo & A. Ganswindt. 2012a. Behavioural inventory of the giraffe (*Giraffa camelopardalis*). *BMC Research Notes* **5**,1: 650–659.

Seeber, P. A., H. T. Ndlovu, P. Duncan, *et al.* 2012b. Grazing behaviour of the giraffe in Hwange National Park, Zimbabwe. *African Journal of Ecology* **50**: 247–250.

Senter, P. 2006. Necks for sex: Sexual selection as an explanation for sauropod dinosaur neck elongation. *Journal of Zoology* **271**: 45–53.

Seymour, R. 2001. Patterns of subspecies diversity in the giraffe, *Giraffa camelopardalis* (L. 1758): Comparison of systematic methods and their implications for conservation policy. PhD Thesis. University of Kent at Canterbury, UK. Chapter 2: The taxonomic status of the giraffe; Chapter 7: Geographic structure in giraffe pelage pattern characteristics.

Seymour, R. 2007. Giraffe taxonomy: Patterns of subspecies diversity in giraffe. *Giraffa* **1**,2: 7–8.

Seymour, R. 2010. More questions than answers: A preliminary interpretation of phylogeographic genetic structure in giraffe. *Giraffa* **4**,1: 28–30.

Seymour, R. 2012. The taxonomic history of giraffe – A brief review. *Giraffa* **6**,1: 5–9.

Shorrocks, B. 2009. The behaviour of reticulated giraffe in the Laikipia district of Kenya. *Giraffa* **3**,1: 22–24.

Shorrocks, B. & D. P. Croft. 2009. Necks and networks: A preliminary study of population structure in the reticulated giraffe (*Giraffa camelopardalis reticulata* de Winton). *African Journal of Ecology* **47**: 374–381.

Shortridge, G. C. 1934. *The Mammals of South West Africa*. Vol. **2**. London: Heinemann.

Sicks, F. 2009. Flesh-coloured tongue in an Angolan giraffe (*Giraffa camelopardalis angolensis*). *Giraffa* **3**,1: 45.

Sicks, F. 2012. Paradoxer Schlaf als Parameter zur Messung der Stressbelastung bei Giraffen (*Giraffa camelopardalis*). PhD thesis. Goethe-Universität, Frankfurt am Main, Germany.

Simmons, R. E. & R. Altwegg. 2010. Necks-for-sex or competing browsers? A critique of ideas on the evolution of giraffe. *Journal of Zoology* **282**: 6–12.

Simmons, R. E. & L. Scheepers. 1996. Winning by a neck: Sexual selection in the evolution of giraffe. *American Naturalist* **148**,5: 771–786.

Sinclair, A.R.E & P. Arcese (eds.). 1995. *Serengeti II: Dynamics, Management, and Conservation of an Ecosystem*. Chicago, IL: University of Chicago Press.

Sinclair, A. R. E., C. Packer, S. A. R. Mduma, *et al.* (eds.). 2008. *Serengeti III: Human Impacts on Ecosystem Dynamics*. Chicago, IL: University of Chicago Press.

Singer, R. & E. Boné. 1960. Modern giraffes and the fossil giraffids of Africa. *South African Museum* **45**: 375–548.

Six, A. & R. Streater-Nunn. 2012. Let me entertain you . . . *Giraffa* **6**,1: 32.

Skinner, J. D. 1966. An appraisal of the eland (*Taurotragus oryx*) for diversifying and improving animal production in southern Africa. *African Wild Life* **20**: 29–40.

Skinner, J. D. & A. J. Hall-Martin. 1975. A note on foetal growth and development of the giraffe *Giraffa camelopardalis giraffa*. *Journal of Zoology, London* **177**: 73–79.

Slijper, E. J. 1946. Comparative biologic-anatomical investigations on the vertebral column and spinal musculature of mammals. *Verh. Kon. Neder. Akad. Wetens., Afd. Naturkunde* **42**: 1–128.

Solounias, N. 1999. The remarkable anatomy of the giraffe's neck. *Journal of Zoology, London* **247**: 257–268.

Sonntag, C. F. 1922. The comparative anatomy of the tongue of the Mammalia. *Proceedings of the Zoological Society of London* **1922**: 639–657.

Spinage, C. A. 1970, Aug 15. Giraffe horns. *Nature* **227**: 735–736.

Stephan, S. A. 1925. Forty years' experience with giraffes in captivity. *Parks and Recreation* **9**: 61–63.

Stott, K. W. & C. J. Selsor. 1981. Further remarks on giraffe intergradation in Kenya and unreported marking variations in reticulated and Masai giraffes. *Mammalia* **45**,2: 261–263.

Strahan, R., P. J. Newman & R. T. Mitchell. 1973. Times of birth of thirty mammal species, bred in the zoos of London and Sydney. *International Zoo Yearbook* **13**: 384–386.

Stratford, K. & S. Stratford. 2011. Using camera traps to investigate drinking patterns in giraffe (*Giraffa camelopardalis*) on Ongava Game Reserve, Namibia. *Giraffa* **5**,1: 30.

Strauss, M. K. L. 2009. Illegal hunting of giraffes: News from northern Tanzania. *Giraffa* **3**,1: 6–7.

Strauss, M. K. L. 2010. Update from the Serengeti Project. *Giraffa* **4**,1: 37.

Strauss, M. K. L. & M. Kilewo. 2011. Snare poaching in the Serengeti: Implications for giraffe population dynamics. *Giraffa* **5**,1: 26.

Strauss, M. K. L. & Z. Muller. 2013. Giraffe mothers in East Africa linger for days near the remains of their dead calves. *African Journal of Ecology* **5**,3: 506–509.

Strauss, M. K. L. & C. Packer. 2013. Using claw marks to study lion predation on giraffes of the Serengeti. *Journal of Zoology, London* **289**,2: 134–142.

Street, P. 1956. *The London Zoo*. London: Odhams Press.

Stutzman, L. & E. Flesch. 2010. Evaluating the dynamics of Thornicrofti's giraffe through photographic database construction. *Giraffa* **4**,1: 15–19.

Sullivan, K., E. van Heugten, K. van Heugten, *et al*. 2010. Analysis of nutrient concentrations in the diet, serum, and urine of giraffe from surveyed North American zoological institutions. *Zoo Biology* **29**: 457–469.

Suraud, J.-P. 2006. The giraffes of Niger are the last in all West Africa. *Giraffa* **1**,1: 8–9.

Suraud, J.-P. 2008. Giraffes of Niger, 2007 census and perspectives. *Giraffa* **2**,1: 4–7.

Suraud, J.-P. 2009. 2008 giraffes in Niger! *Giraffa* **3**,1: 32–33.

Suraud, J.-P. 2011. Identifier les contraintes pour la conservation des dernières girafes de l'Afrique de l'Ouest: Déterminants de la dynamique de la population et patron d'occupation spatiale. PhD thesis. L'Université Claude Bernard – Lyon, France.

Suraud, J.-P., J. Fennessy, E. Bonnaud, *et al*. 2012. Higher than expected growth rate of the endangered West African giraffe *Giraffa camelopardalis peralta*: A successful human–wildlife cohabitation. *Oryx* **46**,4: 577–583.

Suraud, J.-P. & P. Gay. 2010. Giraffe of Niger conservation. *Giraffa* **4**,1: 47–48.

Switek, B. 2009. The ins and outs of the *Sivatherium* snout. *Giraffa* **3**,2: 2–5.

Tarou, L. R., M. J. Bashaw & T. L. Maple. 2000. Social attachment in giraffe: Response to social separation. *Zoo Biology* **19**: 41–51.

Tarou, L. R., M. J. Bashaw & T. L. Maple. 2003. Failure of a chemical spray to significantly reduce stereotypic licking in a captive giraffe. *Zoo Biology* **22**: 601–607.

Taylor, C. R. & C. P. Lyman. 1967. A comparative study of East African antelope, the eland and the Hereford steer. *Physiological Zoology* **40**: 280–295.

Theiler, A., H. H. Green & P. J. du Toit. 1924. Phosphorus in the livestock industry. *South African Department of Agriculture Journal* **8**: 460–504.

Theiler, G. 1962. *The Ixodoidea Parasites of Vertebrates in Africa South of the Sahara (Ethiopian Region)*. Report to Director of Veterinary Services, Onderstepoort. Project S9958. 26 pp.

Thomas, O. 1898. On a new subspecies of giraffe from Nigeria. *Proceedings of the Zoological Society of London* **1898**: 39–41.

Thomas, O. 1901. On a five-horned giraffe obtained by Sir Harry Johnston near Mt Elgon. *Proceedings of the Zoological Society of London* **1901**: 474–483.

Thüroff, J. W., W. Hort & H. Lichti. 1984. Diameter of coronary arteries in 36 species of mammalian from mouse to giraffe. *Basic Research in Cardiology* **79**: 199–206.

Tobler, I. & B. Schwierin. 1996. Behavioural sleep in the giraffe (*Giraffa camelopardalis*) in a zoological garden. *Journal of Sleep Research* **5**: 21–32.

Tutchings, A. 2011. Giraffe indaba – A world first held in Namibia. *Giraffa* **5**,1: 2–3.

Tutchings, A. 2012. Research with borders. *Giraffa* **6**,2: 4.

Urbain, A., J. Nouvel & P. Bullier. 1944. Neóformations cutanées et osseuses de la tête chez les girafes. *Bulletin du Musée National de Histoire Naturelle* Ser 2 **16**: 91–95.

Valdes, E. V. & M. Schlegel. 2012. Advances in giraffe nutrition. In R. Miller & M. Fowler, eds. *Fowler's Zoo and Wild Animal Medicine: Current Therapy*, pp. 612–618. St. Louis, MO: Elsevier.

van Citters, R. L., W. S. Kemper & D. L. Franklin. 1966. Blood pressure responses of wild giraffes studied by radio telemetry. *Science* **152**: 384–386.

van Citters, R. L., W. S. Kemper & D. L. Franklin. 1968. Blood flow and pressure in the giraffe carotid artery. *Comparative Biochemistry and Physiology* **24**: 1035–1042.

van der Jeugd, H. P. & H. H. T. Prins. 2000. Movements and group structure of giraffe (*Giraffa camelopardalis*) in Lake Manyara National Park, Tanzania. *Journal of Zoology, London* **251**: 15–21.

van der Schijff, H. P. 1959. Weidingsmoontlikhede en weideingsprobleme in die Nasionale Krugerwildtuin. *Koedoe* **2**: 96–127.

van Hoven, W. & E. A. Boomker. 1985. Digestion. In R. J. Hudson & R. G. White, eds. *Bioenergetics of Wild Herbivores*, pp. 103–120. Boca Raton, FL: CRC Press.

van Schalkwyk, O. L., J. D. Skinner & G. Mitchell. 2004. A comparison of the bone density and morphology of giraffe (*Giraffa camelopardalis*) and buffalo (*Syncerus caffer*) skeletons. *Journal of Zoology, London* **264**: 307–315.

van Sittert, S. J., J. D. Skinner & G. Mitchell. 2010. From fetus to adult – An allometric analysis of the giraffe vertebral column. *Journal of Experimental Zoology* **314B**: 469–479.

Velhankar, D. P., V. B. Hukeri, B. R. Deshpande, *et al.* 1973. Biometry of the genitalia and the spermatozoa of a male giraffe. *Indian Veterinary Journal* **50**: 789–792.

Vermeesch, J. R., W. DeMeurichy, H. Van Den Berghe, *et al.* 1996. Differences in the distribution and nature of the interstitial telomeric (TTAGGG)(n) sequences in the chromosomes of the Giraffidae, okapi (*Okapia johnstoni*), and giraffe (*Giraffa camelopardalis*): Evidence for ancestral telomeres at the okapi polymorphic rob(4;26) fusion site. *Cytogenetics and Cell Genetics* **72**,4: 310–315.

Verschuren, J. 1958. Ecologie et biologie des grands mammifères (Primates, Carnivores, Ongulés). Bruxelles. *Exploration du Parc National de la Garamba* **9**: 167–172.

Vogelnest, L. & H. K. Ralph. 1997. Chemical immobilisation of giraffe to facilitate short procedures. *Australian Veterinary Journal* **75**,3: 180–182.

von Muggenthaler, E., C. Baes, D. Hill, *et al.* 1999. Infrasound and low frequency vocalizations from giraffe: Helmholtz resonance in biology. *Acoustical Society of America Conference 2001, Riverbanks Consortium.* http://www.animalvoice.com/giraffe.htm

Vosloo, W., S. P. Swanepoel, M. Bauman, *et al.* 2011. Experimental infection of giraffe (*Giraffa camelopardalis*) with SAT-1 and SAT-2 foot-and-mouth disease virus. *Transboundary and Emerging Diseases* **58**: 173–178.

Wakuri, H. & H. Hori. 1970. On the gall bladder of a giraffe (in Japanese). *Journal of the Mammalian Society of Japan* **5**: 41–44.

Walker, J. B. 1974. *The Ixodid Ticks of Kenya*. London: Commonwealth Institution of Entomology.

Walker, J. B. & B. R. Laurence. 1973. *Margaropus wileyi* sp. nov. (Ixodoidea, Ixodidae), a new species of tick from the reticulated giraffe. *Onderstepoort Journal of Veterinary Research* **40**: 13–21.

Wallace, C. & N. Fairall. 1965. Chromosome analysis in the Kruger National Park with special reference to the chromosomes of the giraffe (*Giraffa camelopardalis giraffa*) (Boddaert). *Koedoe* **8**: 97–103.

Warren, J. V. 1974, Nov. The physiology of the giraffe. *Scientific American* **231**,5: 96–105.

Webb, T. D. 2008. Hand-rearing a giraffe (*Giraffa camelopardalis reticulata*) at Miami Metro Zoo. *Giraffa* **2**: 15–17.

Western, D. 1971. Giraffe chewing a Grant's gazelle carcass. *East African Wildlife Journal* **9**: 156–157.

Western, D., S. Russell & I. Cuthill. 2009. The status of wildlife in protected areas compared to non-protected areas of Kenya. *PloS ONE* **4**,7.

Weyrauch, D. 1974. Über das Vorkommen von Parenchymzellteilen im Sinusoidsystem, in sub-endothelialen und interstitiellen Raum der Nebennierenrinde der Masaigiraffe (*Giraffa camelopardalis*). *Anatomischer Anzeiger* **135**,3: 267–276.

Willemse, M. C. A. 1950. The shifting of the molar row with regard to the orbit in *Equus* and *Giraffa*. *Zoologische Mededelingen* **30**: 311–326.

Williams, D. C., P. J. Murison & C. L. Hill. 2007. Dystocia in a Rothschild giraffe leading to a caesarean section. *Journal of Veterinary Medicine* **55**: 199–202.

Wilson, V. J. 1969. The large mammals of the Matopos National Park. *Arnoldia (Rhodesia)* **4**,13: 1–18.

Winton, W. E. de. 1899. On the giraffe of Somaliland (*Giraffa camelopardalis reticulata*). *Annals of the Magazine of Natural History, 7th Series* **4**: 211–212.

Woc-Colburn, M., S. Murray, N. Boedeker, *et al.* 2010. Embryonal rhabdomyosarcoma in a Rothschild's giraffe (*Giraffa camelopardalis rothschildi*). *Journal of Zoo and Wildlife Medicine* **41**,4: 717–720.

Wolfe, B. A., K. K. Sladky & M. R. Loomis. 2000. Obstructive urolithiasis in a reticulated giraffe (*Giraffa camelopardalis reticulata*). *Veterinary Record* **146**, 9: 260–261.

Wood, W. F. & P. J. Weldon. 2002. The scent of the reticulated giraffe (*Giraffa camelopardalis reticulata*). *Biochemical Systematics and Ecology* **30**: 913–917.

Wrobel, K. H. 1965. Das Nierenbecken der Giraffe. *Zeitschrift für Säugetierkunde* **30**: 233–241.

Wyatt, J. R. 1969. The feeding ecology of giraffe (*Giraffa camelopardalis* Linnaeus) in Nairobi National Park, and the effect of browsing on their main food plants. MSc thesis. University of East Africa, Nairobi Kenya.

Wyatt, J. R. 1971. Osteophagia in Masai giraffe. *East African Wildlife Journal* **9**: 157.

Yates, B. C., E. O. Espinoza & B. W. Baker. 2010. Forensic species identification of elephant (Elephantidae) and giraffe (Giraffidae) tail hair using light microscopy. *Forensic Science Medical Pathology* **6**: 165–171.

Yeoman, G. H. & J. B. Walker. 1967. *The Ixodid Ticks of Tanzania*. London: Commonwealth Institution Entomology.

Yong, Z. 1994. Symmetry and other geometric constraints of surface networks in nature and science. *International Journal of Solids and Structures* **32**,2: 173–201.

Young, T. P. & L. A. Isbell. 1991. Sex differences in giraffe feeding ecology – Energetic and social constraints. *Ethology* **87**, 1–2: 79–89.

Ziccardi, F. 1960. The unmaned zebra of Jubaland. *African Wild Life* **14**: 7–12.

Zinn, A. D., D. Ward & K. Kirkman. 2007. Inducible defences in *Acacia sieberiana* in response to giraffe browsing. *African Journal of Range and Forage Science* **24**,3: 123–129.

Index

CPSIA information can be obtained
at www.ICGtesting.com
Printed in the USA
FSHW021020101219
64929FS